Innovative Mechanical Engineering Technologies, Equipment and Materials 2013

IOP Conference Series: Materials Science and Engineering Volume 69

Kazan, Russia
25 – 27 September 2013

ISBN: 978-1-63439-880-0
ISSN: 1757-8981

Innovative Mechanical Engineering Technologies, Equipment and Materials 2013

IOP Conference Series: Materials Science and Engineering Volume 69

Kazan, Russia
25 – 27 September 2013

ISBN: 978-1-63439-880-0
ISSN: 1757-8981

IOP Institute of Physics

Printed from e-media with permission by:

Curran Associates, Inc.
57 Morehouse Lane
Red Hook, NY 12571

Some format issues inherent in the e-media version may also appear in this print version.

Printed by Curran Associates, Inc. (2015)

Institute of Physics
Dirac House, Temple Back
Bristol BS1 6BE UK

Phone: 44 1 17 929 7481
Fax: 44 1 17 920 0979

techtracking@iop.org

Additional copies of this publication are available from:

Curran Associates, Inc.
57 Morehouse Lane
Red Hook, NY 12571 USA
Phone: 845-758-0400
Fax: 845-758-2634
Email: curran@proceedings.com
Web: www.proceedings.com

TABLE OF CONTENTS

IOP Conf. Series: Materials Science and Engineering **69** (2014) 011001 doi:10.1088/1757-899X/69/1/011001

PREFACE

In the period from 25 to 27 September 2013 the city of Kazan hosted the International Scientific Conference "Innovative mechanical engineering technologies, equipment and materials - 2013" (IRTC "IMETEM - 2013"). The conference was held on the grounds of "Kazanskaya Yarmarka" (Kazan). The conference plenary meeting was held with the participation of the Republic of Tatarstan, breakout sessions, forum "Improving the competitiveness and efficiency of engineering enterprises in the WTO" and a number of round tables. Traditionally, the event was followed by the 13th International specialized exhibition "Engineering. Metalworking. Kazan ", in which were presented the development of innovative enterprises in the interests of the Russian Federation of Industry of Republic of Tatarstan, to support the "Foundation for Assistance to Small Innovative Enterprises in Science and Technology" and the 8th specialized exhibition "TechnoWelding".

Kashapov Nail, D.Sc., professor (Kazan Federal University)

Peer review statement

All papers published in this volume of *IOP Conference Series: Materials Science and Engineering* have been peer reviewed through processes administered by the proceedings Editors. Reviews were conducted by expert referees to the professional and scientific standards expected of a proceedings journal published by IOP Publishing.

Modeling the polymer product maceration

D N Ahunov, M N Karpova
Kazan Federal University,
423800, 68/19 Mira avenue, Naberezhnye Chelny, Russia

E-mail: chelny@kpfu.ru

Abstract. The article contains a view of mass transmission simulation procedure conformably to control of manufacturing method's automation, and also is shown a simulator of polymer product maceration process, and results of developed for this simulator realization program system

1. Introduction

Modern technologies for polymeric products producing and processing include heat mass exchange procedures. [1,2]. Due to the complexity of modeling and analysis of such procedures using the mathematical model approach is necessary. Taking into account technological challenges, a new generation of computer technology and computer technology, polymers producing processes mathematical characterizations allow you to simulate in full. The process of modernization and improvement of production management polymeric products associated with the necessity of introducing an integrated automation.

2. The polymer product maceration

One of the principal features of the polymer production technology is the plasticizer's application in the stages of preparation and formation of semi-finished items. Plasticizers provide the required plasticity and during subsequent process step are removed. The necessity to remove them from the composition of the polymer affects the entire manufacturing process of polymers in general [3], since the realization implies a complex set of successive heat mass exchange procedures involving, in particular, the removal of the ether, alcohol ousting with water and at list removing water or other types of drying [4].

The removing process of the solvent from the polymer product is responsible and prolonged [3]. In phase of **dry- curing** ester is removed which makes slowing the evaporation of the alcohol by water vapor saturation of the air approximately up to humidity of 70-85%. At the stage of maceration the group alcohol is removed. The maceration is based on mutual diffusion of water and alcohol [5]. The time of dry- curing and maceration is 36-40 hours.

The article considers the second phase of solvent removal process from the cylindrical shaped polymers. When the polymers maceration water is used for solvent extraction - alcohol and partly ether. At the process of polymers' maceration the water is used to remove solvent - alcohol and partly ether. Water is a medium for dissolving the polymer extracted from the element alcohol. Extraction is possible only in the case when the extractable substance's volume fraction in the dissolution medium is lower than the volume fraction of it in substance from which it is extracted. [6].

Extraction is possible only in the case when the extractable substance's volume fraction in the dissolution medium is lower than the volume fraction of it in substance from which it is extracted. The process stops when the extraction equilibrium between the solution of alcohol in the polymeric product and solution in water.

The process stops when the extraction between the solution of alcohol in the polymeric product and solution in water is equaled. [7].

While the process of running these simulations were considered the following characteristics of the process: increasing the volume fraction of water in the alcohol helps to remove the ether in the process of maceration, however, an excessive increase in the volume fraction of the alcohol extraction is a reason other substances and disturbance of the polymeric structure.

Kinetically extraction obeys the laws mass exchange, convection and molecular diffusion and the laws of transfer extractable substances from the solid phase into the liquid phase [8; 9; 10]. Driving force of transferring the target component is the difference in chemical potentials of its phases.

At simulation following assumptions were made: the convective transfer is disparagingly small; polymeric elements' shrinkage is ignored, because the extracted alcohol is substituted with water (main shrinkage of polymers occurs in the phase of dry- curing).

Constructed simulator consist of nonlinear differential equation in partial differential coefficient second-order in cylindrical coordinates, the first of which describes the process of transferring alcohol and the second transfer process water inside the polymeric element:

$$\frac{\partial U_c(r,z,\tau)}{\partial \tau} = \frac{1}{r}\frac{\partial}{\partial r}\left(r a_{mc} \frac{\partial U_c}{\partial r}\right) + \frac{\partial}{\partial z}\left(a_{mc} \frac{\partial U_c}{\partial z}\right) \quad (1)$$

$$\frac{\partial U_v(r,z,\tau)}{\partial \tau} = \frac{1}{r}\frac{\partial}{\partial r}\left(r a_{mv} \frac{\partial U_v}{\partial r}\right) + \frac{\partial}{\partial z}\left(a_{mv} \frac{\partial U_v}{\partial z}\right), \quad (2)$$

где r, z – spatial coordinates of the selected cylindrical coordinate system;
$U_c(r,z)$, ($U_y(r,z)$) –alcohol (water) concentration in the polymeric products;
a_{mc}, a_{mv} - the diffusion coefficient of the alcohol (water).

The diffusion coefficients depend on the temperature and mass content of the current value:

$$a_{mc} = a_{m_{c_0}} U^{n_1} T^{m_1}; \qquad a_{mv} = a_{m_{v_0}} U^{n_2} T^{m_2}, \quad (3)$$

Where T – temperature; $a_{m_{c_0}}$, $a_{m_{v_0}}$ - the self-diffusion coefficients of water and alcohol; n_1, n_2, m_1, m_2 - some constants, which are calculated experimentally.

Differential equations with partial derivatives are generally countless solutions, therefore the formulation of a diffusion model of a porous material extraction grain polymer will be terminated if, in addition to the diffusion equation (1) and (2) will be [11] the condition of uniqueness - the initial and boundary conditions.

At the initial time $\tau = 0$ the concentration distribution of alcohol and water in the polymer have some specified value (final conditions of dry- curing):

$$U_c(r,z,0) = const = Q_l; \quad U_v(r,z,0) = const = Y_l. \quad (4)$$

For the description of maceration the most used are boundary conditions of the first kind.

On the external boundary the conditions must be fulfilled:

$$U_c(R,z,\tau) = U_c(R_k,z,\tau) = c(\tau); U_c(r,\frac{l}{2},\tau) = c(\tau); \qquad \frac{\partial U_c}{\partial z}\Big|_{z=0} = 0;$$

$$U_v(R,z,\tau) = U_v(R_k,z,\tau) = 1 - c(\tau); U_v(r,\frac{l}{2},\tau) = 1 - c(\tau). \quad \frac{\partial U_v}{\partial z}\Big|_{z=0} = 0. \quad (5)$$

The boundary conditions (5) show the border has an alcohol solution in water with a concentration $c(\tau)$, but the concentration of water in the solution is $1 - c(\tau)$.

$c(\tau)$ – the current concentration of alcohol in the pool, which is calculated by the formula:

$$c(\tau) = \frac{(Q_1 - U)m_n N_n}{m_v + (Q_1 - U)m_n N_n(1 - \frac{p_v}{p_c})}, \quad (6)$$

where m_n - mass ПП, kg; m_v - mass of water, kg; U - average integral concentration of alcohol; N_n - number of polymer cylinders in the pool; p_v - water density; p_c - alcohol density.

Alcohol mass in the pool, as well as $\Delta m_{воды}$ – water mass, diffused into the polymer product, we find, on the basis of knowing the solution of the mikrokinetik problem- the change of alcohol and water average integral concentration.

$$U(\tau) = \frac{1}{V}\iiint_V r \cdot U(r,z,\tau)\, dr\, dz. \quad (7)$$

Based on the model was developed numerical method of calculation [8; 11], and implemented a program complex calculation of the kinetics of the process of soaking. The resulting model of the dynamic properties adequate object. The simulation results show the effect on the number of fills, the processing time and the duration of the process from its initial temperature and solvent content. The bath module affects marginally.
[13].

The construction of mathematical models is an effective tool to study process. Simulation is also necessary to optimize processes and to create automated process control systems.

3. References

[1] Romankiv PG 1975 *Mass transfer processes of chemical technology (systems with solid phase)* Leningrad: Khimiya 336

[2] Galyautdinov R T, Kasparov N F, Luchkin G S 2003 *Inzhenerno-Fizicheskii Zhurnal* **75 (5)** 170-173

[3] Ahunov DN 2010 *Software defined physicochemical properties of polymeric compositions in the manufacture of cellulose nitrate* - Electronic Journal "Education and Science Zakamja Tatarstan", Naberezhnye Chelny, INECA, № 18, http://www.nauctat.ru.

[4] Deacons SG 2010 *Optimal design of industrial devices of chemical technologies based on physical and mathematical modeling* Herald Kasane. tehnol. Univ № 4 131-140

[5] Makhmutova DI 2011*Model regeneration process technology components of energetic condensed systems* Herald Kasane. tehnol. Univ. № 13 156 -160

[6] Akselrud GA 1974 *Extraction (system solid - liquid)* Leningrad: Khimiya 256

[7] Romankiv PG 1983 *Extraction of solids* Leningrad: Khimiya 256

[8] Kurochkin MI 1978 *The extraction and solids leaching. (Library of work - Chemist)* Leningrad: Khimiya 72

[9] AV Lykov 1978 *Heat and mass transfer: Reference* Moscow: Energiya 480

[10] Rudobashta SP 1980 *Mass transport systems in the solid phase* Moscow: Khimiya 248

[11] Tikhonov AP 1976 *Equations of Mathematical Physics* Moscow: Nauka 724

[12] Samara AA 2001 *Mathematical modeling. Ideas. Methods . Examples* 2nd ed. - Fizmatlit, Moscow.

[13] Karpova MN 2011 *Mathematical model of mass transfer in polymer products* Herald Kasane. tehnol. Univ. № 21. - With. 215-218

Modernization of clamping unit of draft of endosurgical monopolar forceps

I.D. Akhmetov[1], N F Kashapov[1] and R N Kashapov[1,2]
[1] Engineering Institute, Kazan Federal University, Kazan, 420008, Russia
[2] Laboratory of Radiation Physics, Kazan Physical-Technical Institute of the Kazan Scientific Center of the Russian Academy of Sciences, Kazan, 420029, Russia

E-mail: kashramiln@gmail.com

Abstract. The object of research is the plasma-electrolytic processing, namely the influence of area ratio of the anode and the cathode to the parameters of the process. Defined area ratio anode to cathode, at which the discharge burning occurs on the cathode, on the anode or on the both electrodes simultaneously. Determine the required ratio of the electrodes ($S_1 \geq 2\ S_2$) for discharge burning of the electrode S_2

1. Introduction

One of the factors of successful treatment in medicine is the using of modern technology. Relatively young and highly effective method of treatment is the laparoscopy. The operations on the internal organs of the abdominal cavity is performed through small openings in the body size of 0.5-1.5 cm. [1,2]. By this method achieves a reduction of trauma, minimizes the risk of infection, shortens recovery and improved cosmetic effect [3]. Electronic system used during laparoscopy allows operation visually enlarge its scope, which increases the accuracy operation.

2. Experimental

For laparoscopic surgery is necessary to use a special medical instrument specific designs and sizes. One of the essential instruments during laparoscopy operation are forceps for coagulation of tissue under the influence of the electron current [3]. There are two types of forceps - monopolar and bipolar types. Now widely used monopolar forceps type. The main problem of this type of instrument is the poor insulation of conductive elements and the complexity of manufacturing the fixing unit end thrust. Thus, the aim of this work was to increase the dielectric properties and the simplification of the process of production of medical instruments.Figure 1 shows a diagram of a laparoscopic instrument (Fig. 1a) used in practice and its thrust (Fig. 1b).

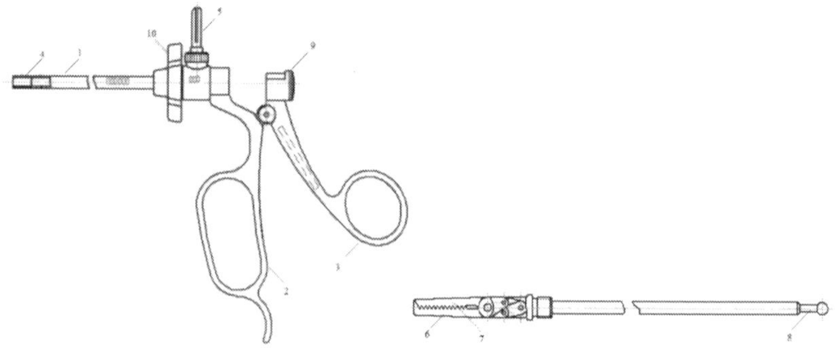

a) b)

Figure 1. The construction of monopolar laparoscopy instrument.

Instrument consists of a shell-tube 1, front immovable handle 2 and rear movable handle 3,function part-forceps 4 and the electrode for current supply 5. The head is designed as an immovable 6 and movable jaw electrodes 7,which pivotally connected with axis.

Fixed jaws connected to the body-tube, which in turn is connected to a fixed handle and ensures the rigidity of the instrument in the longitudinal direction. The movable jaws are electrically connected through the metal of instrument with electrode for current supply and is pivotally connected to the hollow rod 8 extending in the body-tube, interacting with the movable handle through the sleeve 9, which provides electrical insulation. Handles transmit movement of the movable relative to the fixed jaws.

The tool comprises a turning mechanism operating portion 10 about the longitudinal axis of the tool, which comprises a rotating "lamb", is rigidly connected to the housing, and a tube body made of insulating material (material provided NPO "Endomedium"). Having reviewed the construction of forceps, it was found that the insulation properties can be improved by replacing a sleeve movable handle (Fig. 2a) to a standard pen shown in Figure 2b and the end of the movable rod (Figure 2c.) On the dielectric material (Fig. 2c). This is due to the fact that the manufacture of the sleeve requires many operations.

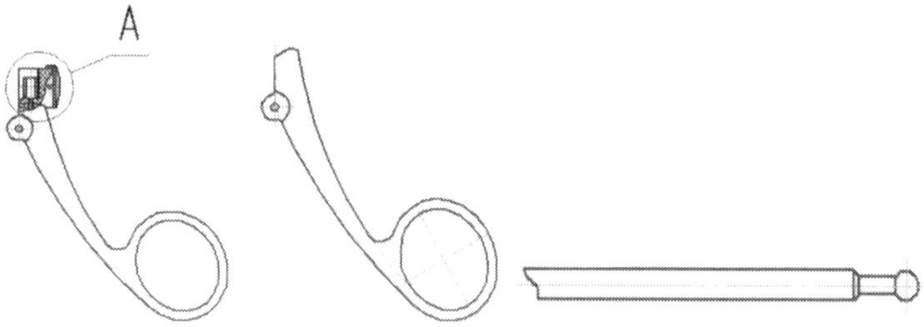

Figure 2. Movable handle with (a),standart handle (b),rod tip (c)

Thus it is necessary to solve two main problems: 1) choose the material has a high dielectric, mechanical strength and thermal properties; 2) to determine the optimal design of the tip and the mechanism of its attachment to the draft. In the design of the tip into account that the instrument must be collapsible, thermally stable, durable and reliable, so all that is necessary during the operation and subsequent cleaning and sterilization. To test was designed three kinds of tips presented in Figure 3. Design of products is carried out in the program KOMPAS 3D. At the output of the model were obtained in "stl" format.

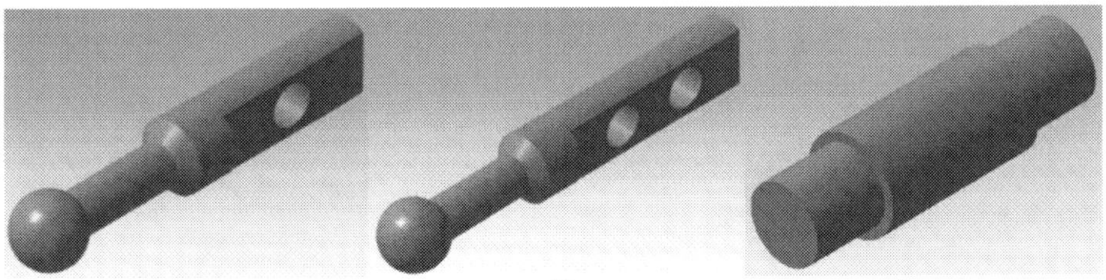

а) б) в)

Figure 3. Model of details

Figure 3a presents a tip overall length of 14 mm and a diameter of 2.5 mm, whose end is a sphere having a diameter of 2.5 mm. Field required for fixing into the sleeve of movable handle. On the opposite side of detail has a hole diameter of 1.5 mm, which is necessary for fixing a metal rod. Figure 3b shows a similar part having a length of 16 mm and two holes for the transverse fixing. Figure 3c product length of 8.4 mm and a maximum diameter of 2.5 mm on both ends of which there are two cylinders with a diameter of 2 mm. This product will be replaced with a metal portion of the thrust on the dielectric, thereby providing electrical protection.

The second important issue is material selection. There are some important requirements to it, namely high flexural strength (750-1000 MPa), the Young's modulus of 200 GPa, fracture toughness 8 MPa m½, the coefficient of thermal expansion of $10 * 10^{-6}$ K^{-1} *, high dielectric properties breakdown voltage 1000 W and with facility in handling.

Such a material is an oxide of zirconium, namely ceramics based on zirconium dioxide ZrO_2 partially stabilized with yttrium oxide (Y2O3) [4]. It possesses all the necessary properties, besides lightness in handling. This can be avoided if the produce is not part of the sintered zirconia strength characteristics when it corresponds to the usual chalk. Also, this material is permitted for medical use, which will significantly simplify the process of launching t in manufacture of the product. [5] In connection with miniaturization of parts of details and the requirements of high accuracy to 10 microns was used for manufacturing CAM-center. Namely, the model developed using CAD-technology machined on an automated 5-axis milling machines Coritec 450i.

This is due to the fact that the CAD / CAM technology has several advantages: high precision, short time production of parts, saving desktop space, automatic trajectory of grinding out, pre-registration material shrinkage during sintering, high performance. [6.7] For generation the g-code program was made by SUM 3D. The place of grinding was chosen with the help of it on your disk and build a strategy of grinding. Since there is grinding of non- sintered zirconium oxide, it is necessary to take into account the shrinkage ratio. In this program, you can pre-enter the shrinkage factor and build a g-code with the required parameters. All this suggests that the most optimal and efficient processing technology zirconia is CAM-technology. Its use eliminates manual work and adjustment on a milling machine, without loss of time allows to adjust the configuration parameters and machining. Upon completion of the program SUM 3D data obtained in the control unit receives a milling machine imes-icore 450i. Time of grinding of three products is 30 minutes.

During the grinding used drills in three sizes: 2.5, 1.02 and 0.6 mm. Drills 2.5 and 1.02 mm are necessary for a rough and medium processing, and 0.6 mm for the final handling. As a material was used standard disc zirconia partially stabilized with yttrium oxide (Y2O3) G 526 with thickness is 12mm and coefficient of shrinkage 1.256. Figure 4a presents photo of blanks of not sintered zirconia after grinding of details. Figure 4b shows after extraction of parts from the workpiece and sintering.

IOP Conf. Series: Materials Science and Engineering **69** (2014) 012002 doi:10.1088/1757-899X/69/1/012002

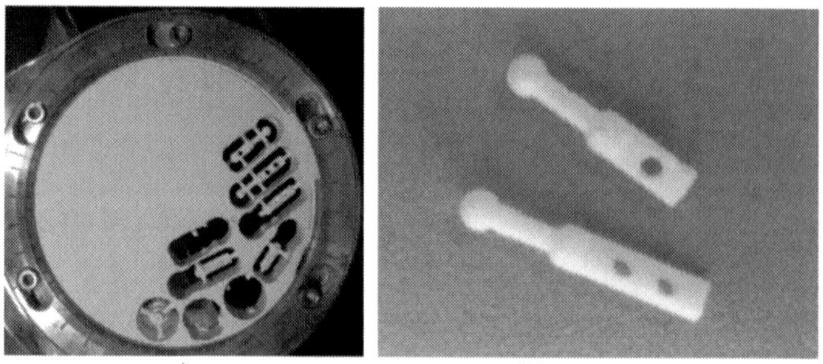

a) b)

Figure 4.Dressed details

Sintering was performed on the controller P 330 firms Nabertherm. For the sintering of details selected sintering regime for small designs. In this mode for 90 minutes, the temperature goes up to 900 ° C. Upon reaching this temperature is holding for 30 minutes, after that for a hundred minutes, the temperature is increased to 1550 ° C. Upon reaching the mark to 1550 ° C temperature rise is stopped and held 2 hours at 1550 ° C. After that produced cooling of products to 200 ° C for 2 hours.

Produced products were tested for dielectric strength and performance characteristics. Tests were conducted on samples using a megger 4102 F / 2-1M at a voltage of 1000V. To this end, replacing the standard dielectric sleeve tool tip, the current value measured in the movable handle. Arrow of megaohmmeter remained at zero. That means no current. Thus was set suitability for the tip of zirconia. Performance tests were carried out by assembling the finished medical instrument. Fixing of metal stainless steel was carried out using the medical cement Fusion Ultra D / C. Figure 5 shows the circuit assembly. Fixing inserts made of zirconium oxide with cement.

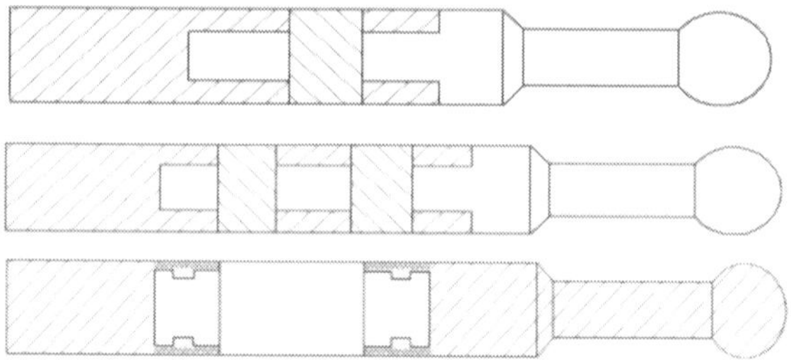

a)

b)

c)

Figure 5 Sceme of assembly tips

The performance characteristics for the three products showed that the first option not satisfied since it is not possible to achieve a strong fixation of the product. The second and third embodiment have been successfully tested on the coagulation process, namely the compression and decompressed of forceps. From the viewpoint of economic efficiency most suitable is third embodiment. From the standpoint of complexity, the second is most effective. As a result of this work we developed three types of thrust tips for endosurgical monopolar forceps. This modification led to the fact that the instrument is much easier to manufacture and more reliable to use. Of the three options, the most appropriate is the second option. This is due to the simplicity and reliability of manufacturing parts.

Acknowledgments

This work was funded by the subsidy allocated to Kazan Federal University for the state assignment in the sphere of scientific activities.

References

1. Dubrovin IA, Klimovich IN, Maskin SS Matyuhin VV Karsanov AM, Yermolayeva NK Laparoscopy in the diagnosis and treatment of postoperative peritonitis // Modern problems of education and science . - 2014. - №1. - p.158

2. Gavrilova, V.A., Kashapov, N.F., Kashapov, R.N. Plasma application of protective polymer-powder coatings to ultrasonic sensors // Biomedical Engineering. - 2011 - 45 (5), pp. 198-200.
3. Sykal NA Programming laparoscopy in the treatment of peritonitis // // Medical emergency conditions. - 2013. - №8 (55). - S.26-28
4.Osiko VV EE Lomonova Polyfunctional materials based on nano-structured crystals of partially stabilized zirconia // Bulletin of the Russian Academy of Sciences. - 2012. - №9. - S.790-800
5.Belov SV, Borik MA, Danileiko YK, Lomonova EE, Osiko VV Riabokon BV, Salyuk VA electrosurgical instruments based on nanostructured crystals dioxide zirconium for dissection and coagulation of tissues // Medical facilities. - 2010. - №4. - S.1-5
6.Andreeva NF, GM Gizetdinova CAD, CAM, CAE, PDM technology // Modern materials, equipment and technology. - 2013. - P. 27-31
7.Tonkonogy VM, Saveliev EV RYAZANTSEV VM, A. Betz Application of CAD / CAM technology in medicine // Proceedings of Odessa Polytechnic University. - 2013. - №1. - S.150-155

Innovative Mechanical Engineering Technologies, Equipment and Materials-2013 IOP Publishing
IOP Conf. Series: Materials Science and Engineering **69** (2014) 012003 doi:10.1088/1757-899X/69/1/012003

Improving of performance system of warranty for automotive engineering abroad on the basis of data of rejections analysis

I V Makarova, R G Habibullin, E I Belyaev, A S Safonov, A A Cheremin
Kazan Federal University,
423800, 68/19 Mira avenue, Naberezhnye Chelny, Russia

E-mail: chelny@kpfu.ru

Abstract. Article deals with improvement possibility of warranty package fulfillment system of automobile vehicles manufacturer in the foreign markets by application of warranty complete sets preparation technique taking into account climatic features of the countries which influence reliability parameters

1. Introduction

Positions' stability of cars producer companies in foreign markets depends on the quality of planning and forecasting, used in the development of the organization's strategy of firm service (FS). Successful development of FS abroad is both in a systematic approach to the creation of both dealer and service network (DSN) and the organization of processes during its operation, which is provided as background information, including the operation of equipment in the region, as well as tools reliability for its analysis and forecasts for subsequent periods.

The purpose of managing of such a system is to meet the customer's applications in a service that depends, particularly, on timely and qualified providing dealership and service center (DSC) by spare parts. The task of delivery planning of spare parts to DSC is mostly solved on an intuitive level. This approach does not lead to losses not only in the absence of relevant items in stock but also in the storage of spare parts unclaimed.

2. The basic part

For more effective planning of structure and spare parts' time of delivery it should be taken into consideration that the various components, accessories and systems in motor vehicles have different resource and have varying degrees of reliability, which, in its turn, depends on many factors that are stochastic.

Reluctance in vehicle occurs at the moment of time $T_{отк}$ that with a certain probability can be predicted. As experience of the technical products' operation shows, changing in the failure rate $\lambda(t)$ of vast majority of objects is described by U – shaped curve and is divided into three operational phases [1 ; 2] in the *break-in* period an increased failure rate is observed, that is connected with running parts and caused usually by manufacturing defects (Im. 1). During the period of full-time operation failures are random and appear suddenly, primarily due to non-compliance with the operating conditions, load changes, the impact of adverse external factors, etc. (Im. 2). The third period is characterized by an increase of the failure rate, which is caused by *aging* and other factors, associated with prolonged use. Considering all above, the provision of spare parts should include functionally distinct mechanisms. Since the *warranty period* is the most important to maintain customer loyalty, firstly the issue of providing quality service exactly in this period is solved.

It's impossible to fight with failures *in running-in period*, but statistical analysis of appeals to auto center helps to reveal details which are mostly susceptible to premature failures. For providing uninterrupted service at the stage of running mechanism calculating the qualitative and quantitative structure of warranty spare parts kits (WSPK) was developed, which are forming for the next batch of sold cars and sent to the operation region along with it.

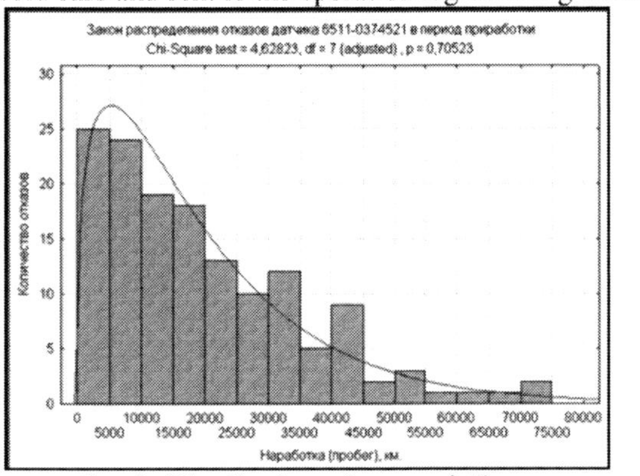

Im. 1. Distribution of sensor failures 6511-0374521 in the break-in period (hidden defects of producer))

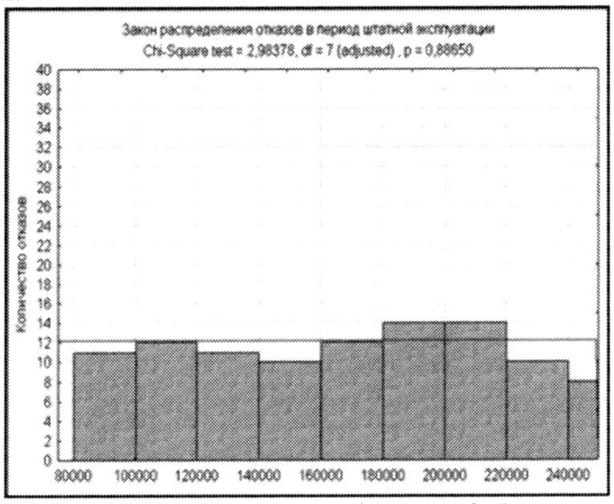

Im. 2. Distribution of sensor failures 6511-0374521 during full-time operation (external influences)

During the *period of full-time operation* failures are mostly depend on operation's conditions, have stochastic character and so the forecast requirements for spare parts is based on the dependencies established by analyzing references in DSC [3].

Assigning the warranty period, the company-producer includes in it not only a break-in period, but part of the full-time operation period. However, as evidenced by the nature of the failures' curve rate, relations of failures number of developments in these areas are described by two fundamentally different laws. Therefore, provision of spare parts during the warranty period should include two functionally distinct mechanisms. In break-in period calculation method of qualitative and quantitative structure of warranty and spare parts kits (WSPK) is required. WSPK for another party of realized cars is formed and sent to the region of operation along with it. At the stage of full-time operation the methodology of possible applications is needed, which will work with failures of this or that detail, taking into consideration the statistics' information data if previous periods. This method of planning is the basis for the formation of structure and calculation of the delivery date in the control center dealer and service network (DSN). The composition of parties in both cases formed the basis of information about failures within the warranty period of operation [4].

As providing of DSN abroad with all necessary spare parts during the warranty period is one of the most important aspects of observation of producer company's warranty liability, the most attention is paid to working the forming methods out. The methodic is formed on the basis of failures' statistic analysis in warranty period. However, in our view, insufficient attention is paid to the account of climatic characteristics of the region.

For confirmation of this thesis the analysis of qualitative and quantitative composition compliance of WSPK for Cuba Republic was made, as it has savannah climate and refers to the second group, selected in accordance with the classification of climates by V.P. Koppen [5].

Information available on the failures has been grouped. Likewise grouping parts that make up WSPK supplied in Cuba Republic were performed. The obtained data were compared, which revealed deviations of WSPK's structure to the number of failures. Such inconsistencies lead to overstocking and freezing of service centers' assets because of a significant proportion of unclaimed parts and, at the same time, lead to decreasing service quality owing to increasing of waiting time because of necessary spare parts lack. During the commercial operation of the vehicle increasing of time being in service leads to loss of customer's confidence in the brand [6 ; 7].

For aligning the structure of WSPK to the number of failures its optimization was performed. Results of redistribution are shown in Table 1, which represents the warranty kit's interest structure with the climatic conditions of the Republic of Cuba.

Table 1.

Comparison of WSPK with the information of failures

Name of detail	Failures (%)	ГКЗЧ before optimization (%)	ГКЗЧ after optimization (%)
Ventilation, heating	0,95	0,16	0,95
Engine	5,69	38,34	6
Wheels and tires	3,89	3,1	3,1
Transmission	2,94	0,16	2,94
Back axle	2,93	4,73	4
Double reduction axle	4,81	4,4	4,9
Window	0,42	0,33	0,33
Forked axle	2,56	3,1	3,1
Bracket	2,12	4,24	2
Tackles	8,7	2,77	8,7
Exhaust system	0,95	2,94	2
Cooling installation	7,06	2,61	7,06
Fuel feed system	5,86	14,68	6
Decelerator system	8,89	5,38	6,85
Engine clutch	8,2	0,98	8,2
Steering system	4,38	0,65	4,38
Cam and roller hoist	9,88	0	9,88
Electrical equipment	19,61	11,26	19,61
TOTAL	100	100	100

As it can be noticed from diagram, given on image 3, optimized WSPK more closely matches the distribution of failures of vehicle components in a given climate group. That's why it can be confirmed that using of a methodology for WSPK, considering the influence of climatic conditions allows to align its structure allocation of vehicle details' failures [8; 9].

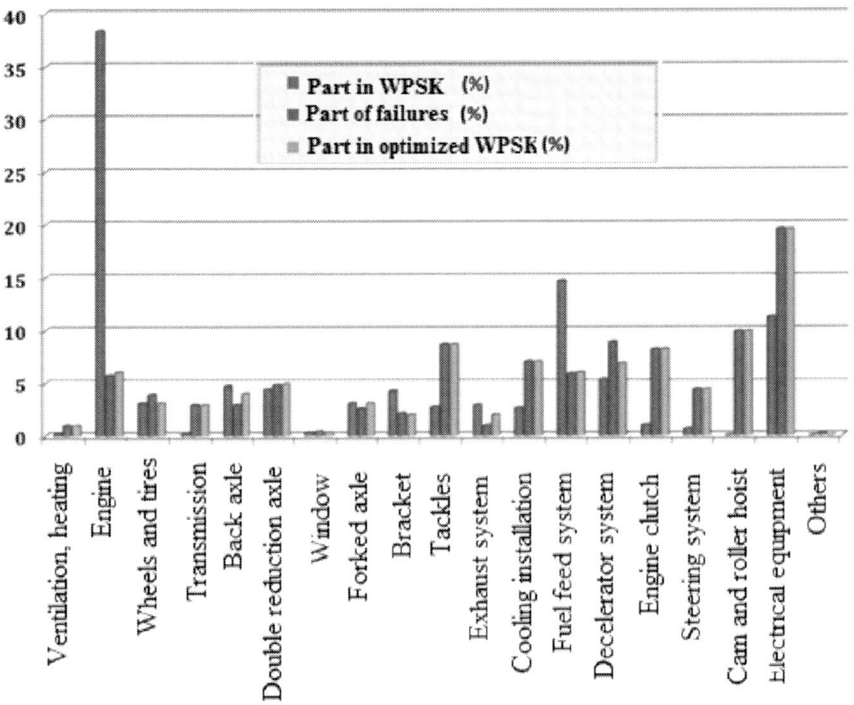

Рис.3. Comparison of WSPK with the information of failures

Annual economic effect for DSS Economic effect of the service center by reducing the unclaimed assets invested in warranty and spare parts kits will be 37,400 rubles. Economic benefits for the car owner, expressed by decrease of lost profits size due to reduced downtime in awaiting repair will be 6992 rubles.

3. References

[1] Malkin V S 2007 *Technical operation of vehicles: the theoretical and the practical aspects.* – M.:Academy – 288 p
[2] Kugel R V 1961 *Durability of vehicles.* "Mashgiz" pub. house
[3] Mirotin L B 2005 *Modern tools of logistics management* – M.: "AST Publishing Group" 494 p
[4] Habibullin R G 2012 *Supply management of spare parts in the system of firm service vehicles by predicting appeals* // Innovative information technology № 1 P 563-565
[5] Koeppen V P 1938 *Fundamentals of climatology (Climats of the globe).* M.: Uchpedgiz – 376 p
[6] Kramarenko G V 1983 Technical operation of cars - M: Transportation. – 488 p
[7] Khalafyan A A 2007 *STATISTICA* 6. Statistical analysis. – M.: Binorm-Press 512 p
[8] Makarova I V 2011 *Application of modern methods of modeling and management for improving the effectiveness of the firm service car* Vestnik of IzhSTU № 1 P 118-121
[9] Habibullin R G 2012 *Supply management of spare parts in the system of firm service vehicles by predicting appeals* // Innovative information technology № 1 P 563-565
[10] Gertsbakh I B 1966 *Failures model* -M.: Soviet radio 166 p

Mathematical model of the component mixture distribution in the molten cast iron during centrifugation (sedimentation)

R A Bikulov, LM Kotlyar
Kazan Federal University,
423800, 68/19 Mira avenue, Naberezhnye Chelny, Russia

E-mail: chelny@kpfu.ru

Abstract. For the development and management of the manufacturing processes of axisymmetric articles with compositional structure by centrifugal casting method [1,2,3,4] is necessary to create a generalized mathematical model of the dynamics of component mixture in the molten cast iron during centrifugation. In article. based on the analysis of the dynamics of two-component mixture at sedimentation, a method of successive approximations to determine the distribution of a multicomponent mixture by centrifugation in a parabolic crucible is developed.

1. Introduction

Establishing regularities of structure formation in cast iron in conditions of forced heat and mass transfer by varying the temperature and time parameters of the melting and crystallization is not only the subject of theoretical interest, but also the basis for the development of new manufacturing processes of axisymmetric products with compositional structure by the method of centrifugal casting by centrifugation of the melt. One of the stages of research includes the development of a mathematical model of the transfer and distribution of mixture components in the iron cast melt during centrifugation.

Lamm equation [6], which describes the process of diffusion and sedimentation of substances have been widely used for the study of macromolecules with the aid of analytical ultracentrifugation [6, 7, 8, 9, 10].

To develop mathematical model for assessing the distribution of components in the molten cast iron during centrifugation process, we should consider the known equations describing the process of transfer of matter in the centrifuge cell with diffusion and sedimentation for 2-component mixture [5].

2. The basic part

We shall assume that at the initial time τ_0 all the mixture components are distributed evenly throughout the volume of the filled crucible V_0, have the same temperature T_0 and the weight of all components is M_0. Let the in the initial time τ_0 mass concentration of the components of mixture components are given $C_k^0(\kappa = \overline{1,n})$, density of all components $\rho_k^0(\kappa = \overline{1,n})$. In general, ρ_κ are functions of temperature T и $\rho_k^0 = \rho_k^0(T_0)$, where T_0 – the temperature of the mixture at $\tau=\tau_0$. Then, the average density of the mixture ρ_0 can be determined by the formula

$$\rho_k^0 = {M_0}/{V_0} \quad , \tag{1}$$

When centrifuging molecules or particles move under the influence of artificial inertial force directed perpendicularly to the axis of rotation.

We assume, following [5] that the mixture consists of two components. As a standard velocity of the interface U, we accept the change of the radial coordinate $dr_{cp}/d\tau$ at the midpoint.

Speed value U can be represented as:

$$U = \frac{dr_{cp}}{d\tau} = \omega^2 rs \,, \tag{2}$$

Where ω - the angular velocity of rotation, rev / minute; r - distance to the axis of rotation, m; s – sedimentation constant characterizing, as it follows from the equation itself, the rate of deposition of particles per unit of centrifugal acceleration $\omega^2 r$. The solution of this equation is easy, if you know the sedimentation constant s.

Unfortunately, this constant must be determined by getting the experimental dependence $\ln\left(r_{cp}(\tau)/r_{cp}(\tau_0)\right)$ of the equation. In the case of multicomponent mixtures more subtle experimental methods are required.

Process of material transportation in the centrifuge is described by Lamm equation [2] which for a radial sector of the cell can be written as:

$$\left(\frac{dC}{d\tau}\right)_r = -\frac{1}{r}\left\{\frac{d}{dr}\left[\omega^2 rs\,C - Dr\left(\frac{dC}{dr}\right)_\tau\right]\right\}_\tau , \tag{3}$$

wherein $\left(\frac{dC}{d\tau}\right)_r, \left(\frac{dC}{dr}\right)_\tau$, - the concentration function derived (kg / l) at constant values r and τ correspondingly, D - the diffusion coefficient [m2/c], s - sedimentation constant characterizing the particle sedimentation rate per unit centrifugal acceleration $\omega^2 r$. This equation, as shown in [6,7] has not, in general, the exact analytical solution. Nevertheless, for two extreme cases there are the exact solutions: when $D = 0$ the (lack of diffusion) and $s = 0$ (no sedimentation). Since we are interested in the case when the effect of diffusion comparing to inertial forces is small, we consider the solution of equation (3) with D = 0.

Taking into account Fick equation, which describes the process of diffusion and sedimentation of interface in an infinite stream in the direction which is perpendicular to the rotation axis OX [6].

When $D = 0$, we obtain $x = x_0 \exp(s\omega^2\tau)$,

x_0 - where - the interface (meniscus), and taking into account (1), we find

$$\left\{\begin{array}{l} C(x,\tau) = 0; \quad \text{при } x \in (x_0, x_\tau) \\ C(x,\tau) = C_0 \exp(-2s\omega^2\tau); \quad \text{при } x \in (x_\tau, x_\partial) \end{array}\right\} . \tag{4}$$

where x_0 - the initial coordinate of mixture interface (meniscus), m;

x_{∂} - coordinate of the crucible bottom, m; C_0 - initial concentration, kg / l.

Distribution of concentration over time in this case, has the plateau with a constant concentration. The essential fact in this solution is that over time the particles of the corresponding component concentrate on the side surface of the crucible with a higher density and the lower density components are distributed along the side surface of the preceding components.

Taking into consideration that the limit solution of Lamm equation (4) and the fact that the assigned task of the components distribution in volume where $\tau \to \infty$ does not require knowledge of the prehistory of the process, you can reduce it to a fairly simple algorithm.

Suppose that the form of the crucible is defined by the equation:

$$y = \frac{H_* x^2}{R_*^2}, \tag{5}$$

where R_*, H_* - constants that are the coordinates of an arbitrary point on the boundary section.

We order the components of the mixture in ascending given when τ_0 densities $\rho_1 < \rho_2^\circ, < \ldots < \rho_n^\circ$ and define the corresponding mass concentration $C_k^\circ \ (k = \overline{1,n})$ of components of the mixture at the initial time τ_0.

While centrifugation component molecules move under the influence of artificial gravitational field and after a certain time, according to a particular solution of the equation Lama (3), will be located in layers.

We assume that the rotational speed of the crucible is bounded both above and below so that the contents of the crucible did not form the gaps near the axis and inertial forces exceed gravity.

To determine the limits of mixture components developed a method of successive approximations.

For the particular case of three-component mixture $Fe-C-Si$ at concentrations equal to C, $Fe-92.45\%$; $C-5.1\%; Si-2.45\%$, total mass $M_0 = 100г$ and a predetermined shape of the crucible H*=12cм, R*=3,6 cm calculations have been made the results of which after three iterations are shown in table 1.

Table 1
Table of calculations results

Coordinates / Iteration №	r_1 (cm)	r_2 (cm)	r_3 (cm)	H_1 (cm)	H_2 (cm)	H_3 (cm)
0	0,692	0,8368	1,8710	2,974	2,974	3,150
1	0,685	0,8283	1,8517	2,986	2,992	3,062
2	0,690	0,8343	1,8651	2,986	2,985	3,018

Here r_k, H_k coordinates of the upper boundary point of "k"-th mixture ($K = \overline{1,3}$).

3. References

[1] Bikulov R A 2009 *Production of multi-purpose cast iron: the development of compositions and process control.* - Moscow: Academia 351 p. ISBN 978-5-87444-285-9
[2] Bikulov R A 2010 *Production of composite materials and cast iron for special purposes for engineering (tutorial)* - M.: Publishing house "paleotypes" 239 p
[3] Samarin AM 1967 *Vestnik AN SSSR. Structure of metallic melts.*, № 7, P 59-67
[4] Azarov A I, Kashapov N F, Osipova O P 2003 *Litejnoe Proizvodstvo.*, №11, p 30-31
[5] Serduk I N, O N Evseeva 2006 *New opportunities for analytical ultracentrifugation analysis of hydrodynamic properties of proteins analysis* / Success of Biological Chemistry - V.46. P 349-372.
6. Lamm, O. (1929) Akr. Mat. Astr.Fys., №21, p 1–2
[7] Fujita H 1962 *Mathematical Theory of Sedimentation Analysis?* New York: Academic Press
[8] Hensley P Structure. 1996, №4, p 367-373
[9] Hansen J C, Lebowitz J, Demeler B 1994 *Biochemistry.*, №33, p 13155-13163
[10] Stafford W F 1994 *Methods Enzymol.*, III, № 240, p 478-501

Numerical experiment optimization to obtain the characteristics of the centrifugal pump steps package

S V Boldyrev, A V Boldyrev
Kazan Federal University,
423800, 68/19 Mira avenue, Naberezhnye Chelny, Russia

E-mail: chelny@kpfu.ru

Abstract. The numerical simulation method of turbulent flow in a running space of the working-stage in a centrifugal pump using the periodicity conditions has been formulated. The proposed method allows calculating the characteristic indices of one pump step at a lower computing resources cost. The comparison of the pump characteristics' calculation results with pilot data has been conducted.

1. Introduction

Centrifugal pump characteristics are received experimentally on special stands [1; 2]. Moreover, several stages (figure 1) are used during testing compound pumps in order to exclude the effect of input and output devices on the results. Although, if there is quality measuring equipment, precise and quick at obtaining characteristics, its use in the design of new constructions leads to the costs brought about by the need of experimental samples refinement.

A numerical modeling of currents flowing in a running space of a pump can be considered to be an alternative way to locate pump characteristics in intermediate stages of design refinement to prototyping. However, its application requires [3; 4; 6; 7]: adequate mathematical model, optimal boundary conditions, quality computational grid with a sufficient number of cells, solver correct parameters. Therewith, the numerical experiment implies the licensed software and ample computational resources (to reduce time costs).

This work is devoted to assessing ways of numerical experiment optimization for the compound pumps.

2. Basic part

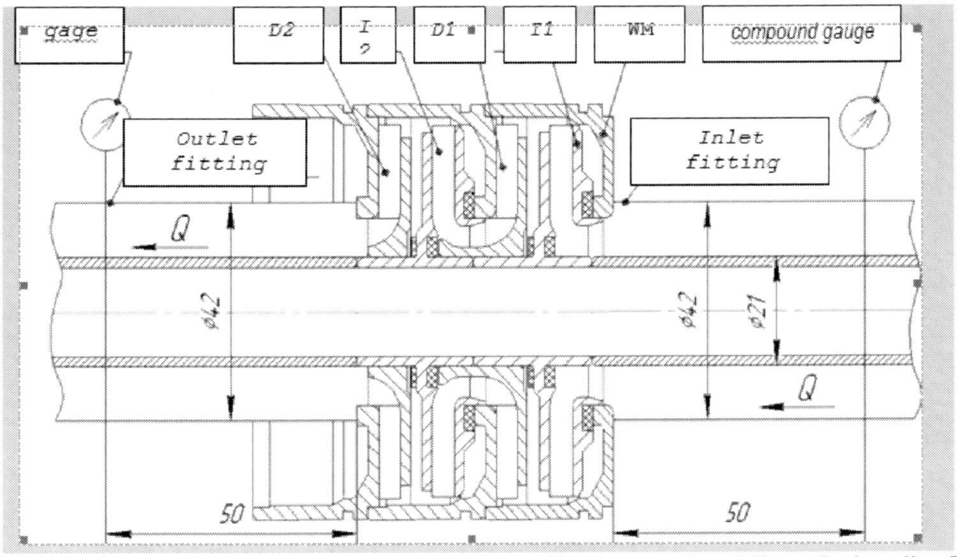

Fig.1. The experimental setup for testing compound centrifugal pump (D – diffuser, I – impeller, WM – wheel mount)

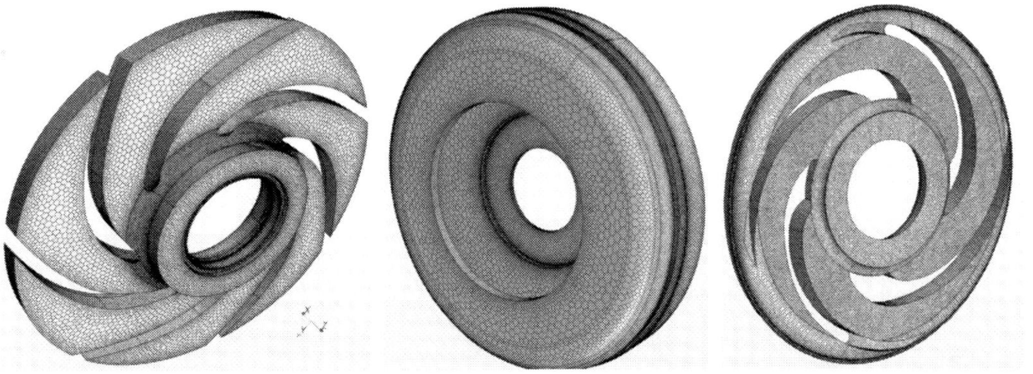

Fig.2. Three-dimensional geometric models and running spaces computational grids at a pump stage: the impeller (left), the intermediate zone (center) and the diffuser (right)

A numerical simulation of turbulent water flow has been performed in the running space of the compound centrifugal pump ETsNA(K)5A-80 in the licensed software package Star-CCM+ [8].

A mathematical model has been composed of Reynolds averaged Navier-Stokes equations and continuity (including stationary and rotating coordinate systems), equations of two-layer Realizable turbulence model $k-\varepsilon$ [5] with hybrid wall functions [9, 10].

Running space of a stage is conditionally divided into 3 areas (figure 2): an impeller (PK), a diffuser (HA), an intermediate area (formed by the outer surfaces of the impeller and a diffuser, besides by and inner surface of the pump casing). A computational grid (figure 2) consisting of many-sided cells, the size of which is chosen to be proportional to its square, is applied for the discretization of the areas. Near walls the grid is crushed with layers of prismatic cells so that the center of wall cells is within a buffer zone [9] of a boundary layer (recommended for the model of turbulence). The total number of control volumes in the grid has come up to 4.5 million.

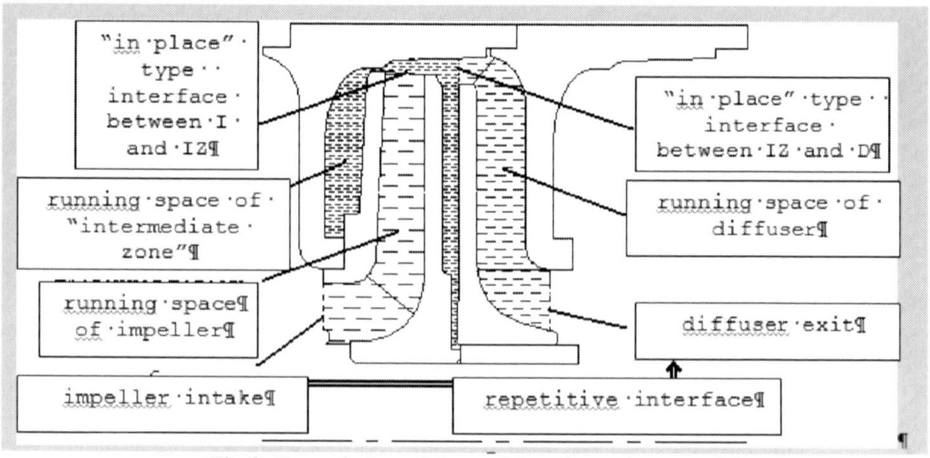

Fig.3. Formulation of the problem (boundary conditions)

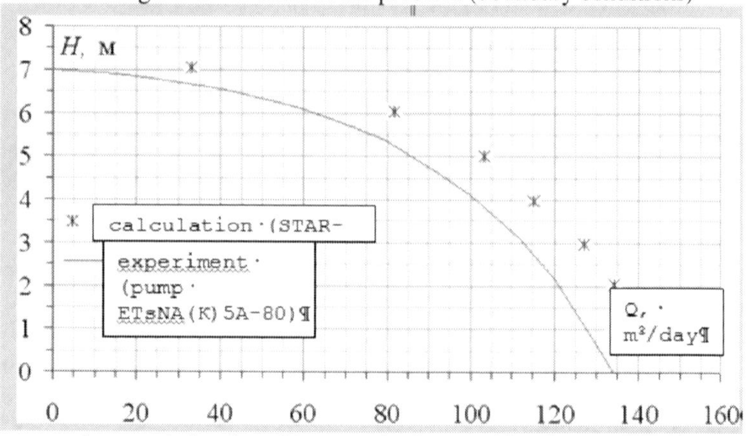

Fig.4. Pressure characteristic of working-step in the centrifugal pump ETsNA(K)5A-80

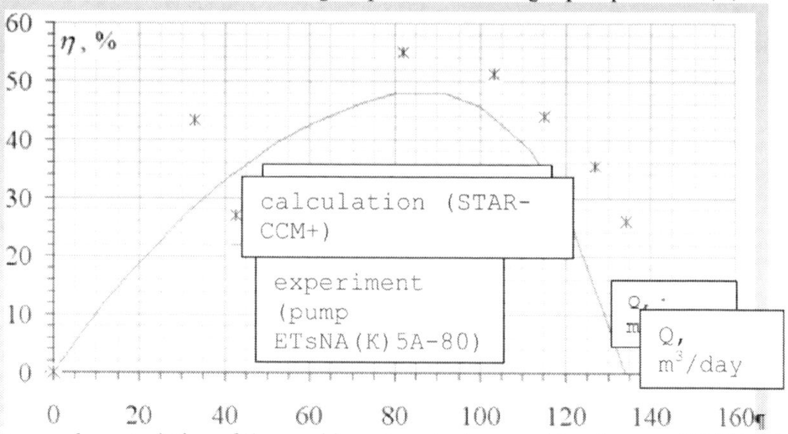

Fig.5. Energy characteristics of the working-stage in the centrifugal pump ETsNA(K)5A-80

There are several variants of boundary conditions in flow modelling within vane pumps [3; 8; 5; 6]:

1) If the input is set to the mass flow rate (corresponding to the flow at the selected point of characteristics) and the output is constant static pressure, in the simulation result is determined by the pressure at the inlet, whereby pressure pump is calculated. If the input is set to a constant total pressure, and the output is the static pressure (corresponding to the pressure at the selected point of

characteristics), pump flow is determined.

2) Three-dimensional running space is treated either in whole or partially (to save computing resources), a part of running space being used around one blade inside each paddle wheel of a pumping step via use of circular frequency currents. Since the result on adjacent borders of computational spaces is "stitched" by making use of special interfaces such as "in place" [8], the last option is a strong fit only when used for the same number of rotor wheels and pump diffusers.

It should be noted that in all above cases, it is necessary to simulate three-dimensional turbulent flow in at least three stages of the pump in order to reduce charge and discharge devices′ impact on characteristics.

The method of setting the boundary conditions, allowing calculating the characteristics of a single-stage pump with a lower cost of computing resources (Fig.4) is presented in this paper. The impeller intake and diffuser exit are connected by the periodic interface, which provides the flow stabilization in the step with the specified constant pressure differential corresponding to the pressure in the selected point of characteristics. The “impermeability” and "nonslip" conditions are assigned on fixed boundaries. The result of simulation is the current magnitude of the pump flow.

The comparison of the calculated characteristics of the stage (Fig.4, 5) with the experimental data [11] allows making the following conclusions:

1) good quality and satisfactory quantitative match of the results with the field stage testing data is obtained;

2) uprating of pump efficiency magnitudes (Fig.5) can be explained by the fact that in the calculations the textolite leathers friction of impellers against the pump diffusers boundaries and the friction in the bearings were not taken into account.

Modern technologies for polymeric products producing and processing include heat mass exchange procedures. [1,2]. Due to the complexity of modeling and analysis of such procedures using the mathematical model approach is necessary. Taking into account technological challenges, a new generation of computer technology and computer technology, polymers producing processes mathematical characterizations allow you to simulate in full. The process of modernization and improvement of production management polymeric products associated with the necessity of introducing an integrated automation.

2. The polymer product maceration

One of the principal features of the polymer production technology is the plasticizer’s application in the stages of preparation and formation of semi-finished items. Plasticizers provide the required plasticity and during subsequent process step are removed. The necessity to remove them from the composition of the polymer affects the entire manufacturing process of polymers in general [3], since the realization implies a complex set of successive heat mass exchange procedures involving, in particular, the removal of the ether, alcohol ousting with water and at list removing water or other types of drying [4].

The removing process of the solvent from the polymer product is responsible and prolonged [3]. In phase of **dry- curing** ester is removed which makes slowing the evaporation of the alcohol by water vapor saturation of the air approximately up to humidity of 70-85%. At the stage of maceration the group alcohol is removed. The maceration is based on mutual diffusion of water and alcohol [5]. The time of dry- curing and maceration is 36-40 hours.

The article considers the second phase of solvent removal process from the cylindrical shaped polymers. When the polymers maceration water is used for solvent extraction - alcohol and partly ether. At the process of polymers' maceration the water is used to remove solvent - alcohol and partly ether. Water is a medium for dissolving the polymer extracted from the element alcohol. Extraction is possible only in the case when the extractable substance's volume fraction in the dissolution medium is lower than the volume fraction of it in substance from which it is extracted. [6].

IOP Conf. Series: Materials Science and Engineering **69** (2014) 012005 doi:10.1088/1757-899X/69/1/012005

Extraction is possible only in the case when the extractable substance's volume fraction in the dissolution medium is lower than the volume fraction of it in substance from which it is extracted. The process stops when the extraction equilibrium between the solution of alcohol in the polymeric product and solution in water.

The process stops when the extraction between the solution of alcohol in the polymeric product and solution in water is equaled. [7].

While the process of running these simulations were considered the following characteristics of the process: increasing the volume fraction of water in the alcohol helps to remove the ether in the process of maceration, however, an excessive increase in the volume fraction of the alcohol extraction is a reason other substances and disturbance of the polymeric structure.

Kinetically extraction obeys the laws mass exchange, convection and molecular diffusion and the laws of transfer extractable substances from the solid phase into the liquid phase [8; 9; 10]. Driving force of transferring the target component is the difference in chemical potentials of its phases.

At simulation following assumptions were made: the convective transfer is disparagingly small; polymeric elements' shrinkage is ignored, because the extracted alcohol is substituted with water (main shrinkage of polymers occurs in the phase of dry- curing).

Constructed simulator consist of nonlinear differential equation in partial differential coefficient second-order in cylindrical coordinates, the first of which describes the process of transferring alcohol and the second transfer process water inside the polymeric element:

$$\frac{\partial U_c(r,z,\tau)}{\partial \tau} = \frac{1}{r}\frac{\partial}{\partial r}\left(r a_{mc}\frac{\partial U_c}{\partial r}\right) + \frac{\partial}{\partial z}\left(a_{mc}\frac{\partial U_c}{\partial z}\right) \qquad (1)$$

$$\frac{\partial U_v(r,z,\tau)}{\partial \tau} = \frac{1}{r}\frac{\partial}{\partial r}\left(r a_{mv}\frac{\partial U_v}{\partial r}\right) + \frac{\partial}{\partial z}\left(a_{mv}\frac{\partial U_v}{\partial z}\right), \qquad (2)$$

где r, z – spatial coordinates of the selected cylindrical coordinate system;
$U_c(r,z)$, $(U_y(r,z))$ –alcohol (water) concentration in the polymeric products;
a_{mc}, a_{mv} - the diffusion coefficient of the alcohol (water).

The diffusion coefficients depend on the temperature and mass content of the current value:

$$a_{mc} = a_{m_{c_0}} U^{n_1} T^{m_1}; \qquad a_{mv} = a_{m_{v_0}} U^{n_2} T^{m_2}, \qquad (3)$$

Where T – temperature; $a_{m_{c_0}}$, $a_{m_{v_0}}$ - the self-diffusion coefficients of water and alcohol; n_1, n_2, m_1, m_2 - some constants, which are calculated experimentally.

Differential equations with partial derivatives are generally countless solutions, therefore the formulation of a diffusion model of a porous material extraction grain polymer will be terminated if, in addition to the diffusion equation (1) and (2) will be [11] the condition of uniqueness - the initial and boundary conditions.

At the initial time $\tau = 0$ the concentration distribution of alcohol and water in the polymer have some specified value (final conditions of dry- curing):

$$U_c(r,z,0) = const = Q_l; \quad U_v(r,z,0) = const = Y_l. \qquad (4)$$

For the description of maceration the most used are boundary conditions of the first kind.

On the external boundary the conditions must be fulfilled:

$$U_c(R,z,\tau) = U_c(R_k,z,\tau) = c(\tau(\tau U_c(r,\frac{l}{2},\tau) = c(\tau(\tau \qquad \frac{\partial U_c}{\partial z}\Big|_{z=0} = 0;$$

$$U_v(R,z,\tau) = U_v(R_k,z,\tau) = 1 - c(\tau(\tau U_v(r,\frac{l}{2},\tau) = 1 - c(\tau(. \frac{\partial U_v}{\partial z}\Big|_{z=0} = 0. \qquad (5)$$

The boundary conditions (5) show the border has an alcohol solution in water with a concentration $c(\tau($, but the concentration of water in the solution is $1 - c(\tau)$.

$c(\tau)$ – the current concentration of alcohol in the pool, which is calculated by the formula:

$$c(\tau) = \frac{(Q_1 - \overline{U})m_n N_n}{m_v + (Q_1 - \overline{U})m_n N_n (1 - \frac{p_v}{p_c})}, \tag{6}$$

where m_n - mass ПП, kg; m_v - mass of water, kg; $\overline{U}$ - average integral concentration of alcohol; N_n - number of polymer cylinders in the pool; p_v - water density; p_c - alcohol density.

Alcohol mass in the pool, as well as $\Delta m_{воды}$ – water mass, diffused into the polymer product, we find, on the basis of knowing the solution of the mikrokinetik problem- the change of alcohol and water average integral concentration.

$$\overline{U}(\tau) = \frac{1}{V}\iiint_V r \cdot U(r,z,\tau)\, dr\, dz. \tag{7}$$

Based on the model was developed numerical method of calculation [8; 11], and implemented a program complex calculation of the kinetics of the process of soaking. The resulting model of the dynamic properties adequate object. The simulation results show the effect on the number of fills, the processing time and the duration of the process from its initial temperature and solvent content. The bath module affects marginally. [13].

The construction of mathematical models is an effective tool to study process. Simulation is also necessary to optimize processes and to create automated process control systems.

3. References

[1]. GOST 6134-2007 (ISO 9906:1999). *Dynamic pumps.* Test methods.

[2]. Sherstyuk A N *Pumps, fans and compressors: textbook for colleges.* 1972 Moscow,344 pp.

[3]. Zharkovsky A A *Simulation of viscous flow in the centrifugal pump impeller* 2011 Compressors and pneumatics. №4. – P.18.

[4]. Slobodkina FA *Fluid flow in centrifugal pump step* 2008 Mathematical modeling №10. – Pp. 51-63.

[5]. Shih T.-H., Liou, W W, Shabbir A, Yang Z. and Zhu J. 1994. *A New k-□ Eddy Viscosity Model for High Reynolds Number Turbulent Flows – Model Development and Validation,* NASA TM 106721.

[6]. Lomakin V O *Verification of calculation results in the computational fluid dynamics package STAR-CCM+ flowing part of a centrifugal pump AKh 50-32-200* / V O Lomakin, A I Petrov // Journal of higher schools. Engineering., 2012. – №5. – Pp. 6-9.

[7]. Sheypak A A *Experimental and computational study of the centrifugal pump characteristics* / A. A. Sheypak, Ya. K. Lokhansky, A. S. Pozdnyakov // Mechanical industry and engineering education. – 2009. – № 1. – Pp. 53-61.

[8]. User Guide STAR-CCM+ Version 7.04.006. - CD-adapco, 2012.

[9]. Loitsyanskii L G *Mechanics of fluid and gas.*2003 – 7-th ed., corr. – Moscow, 846 pp.

[10]. Abdullin I Sh, Galyautdinov R T and Kashapov N F *Inzhenerno-Fizicheskii Zhurnal* 2004**74 (5)** 104-107

[11]. *Installation of submersible centrifugal pumps (UETsNA) in full set: serial production catalogue.*2008 JSC "ALNAS".

IOP Conf. Series: Materials Science and Engineering **69** (2014) 012006 doi:10.1088/1757-899X/69/1/012006

Experimental installation to study the influence of acoustic oscillations on the characteristics of a glow discharge

S A Fadeev, N F Kashapov and V M Larionov
Kazan Federal University, 18 Kremlyovskaya str., Kazan 420008, Russian Federation

E-mail: Sergey_Fadeev@kpfu.ru

Abstract. It is proposed a sketchy scheme of experimental installation to study the influence of acoustic oscillations having longitudinal, tangential, radial and combined forms on the parameters of a glow discharge.

1. Introduction

For half a century the development of laser technology have been created various types of lasers are designed for a variety of scientific and engineering problems. The effective use of laser technology are sufficiently diverse – this material processing, communication, computer science, medicine and many others.

The most significant area of application of lasers is laser processing of materials, which is based, in most cases on the thermal effect of laser radiation. The advantages of laser machining in comparison with other methods is chemical purity, controlled temperature profile, the shape and depth of the zone of thermal effects, small additional machining, contactless operation and ease of automation [1].

Among laser sources used for the treatment of materials, the most widely used high-power gas (CO_2) lasers. Thereby actual task is to develop new ways to increase the power of industrial lasers to produce a significantly higher energy efficiency to make significant progress in the development process of laser treatment.

Our research team carries out studies in the field of vacuum-plasma and plasma-chemical processes production of functional coatings and products. In [2] presents the rationale of effective influence of the transverse acoustic oscillations on constricted electrical discharge that leads to the preservation of the diffuse form of the glow discharge and increase the energy input to the discharge of a CO_2 laser. Objects of research in [3] are coatings and technology of their applying to the piezoelectric elements for ultrasound. Method of plasma coating to restore the defective medical ultrasonic sensors is presented in [4]. In [5] investigated non-independent electrical discharge in the longitudinal flow of dust-electron plasma. In this paper we propose a schematic diagram of the experimental setup to investigate the influence of different forms of acoustic oscillations on the parameters of the glow discharge. Using this experimental setup we can test and debug an acoustic method of preventing instabilities.

2. Methods of prevent the occurrence of instabilities

Uniform diffuse state of the positive column of a glow discharge in the CO_2 laser is often unstable, especially when the discharge is in large volumes at high pressures, high current, and Joule heat release. Random perturbations increase and the plasma goes into a different spatially inhomogeneous state – contraction. Contraction – compression of the plasma in a brightly glowing cord, and then goes into the arc. Development of the instability of this type in the electric discharge chamber of gas lasers stops lasing laser radiation. Overcoming the contraction of the discharge is the most difficult task in the development of high-power gas lasers.

There are several traditional ways to overcome the instabilities – partitioning the cathode, pumping gas through the discharge gap, the excitation of small-scale turbulence in the gas flow, the use of self-sustaining discharge. Integrated use of these methods can significantly increase the energy input into the diffuse discharge [6, 7].

Increased interest is the acoustic method prevent the occurrence of instabilities glow. Resonance phenomenon of acoustic oscillations in a confined space is widely studied and used in various types of devices: the combustion chambers of rocket and aircraft engines, and other heat power installations. We should also mention vibration combustion, which is accompanied by an acceleration of heat transfer to the walls of the combustion chamber, improving combustion efficiency in comparison with the stable stationary combustion mode [8, 9].

The essence of the method is to prevent the occurrence of instabilities of the glow discharge consists in excited by acoustic oscillations in the discharge chamber of a CO_2 laser. The presence of acoustic oscillations in the discharge gap is accompanied by a modification of a bell-shaped spatial distribution of the electron density along the radius of the tube on the parabolic. Increases the longitudinal electric field decreases the radial temperature gradient, increased diffusion losses of charged particles on the wall [10]. Displaying when the sound is muted a inverted population of the gas in the tube with the subsequent generation of flash super luminescent light, supporting the homogeneous state of the excited gas, due to a series of recombination and relaxation processes in the plasma [11].

3. The apparatus of the experimental installation

As shown in [2, 12] the excitation of the transverse acoustic oscillations in the discharge chamber of CO_2-lasers could improve the acoustic method for preventing the occurrence of instabilities. But before apply this technology in the development of CO_2-lasers, it is necessary to conduct comprehensive investigations of the influence of the transverse acoustic oscillations on the glow discharge. For this purpose developed an experimental installation for the studies of the influence of acoustic oscillations on the characteristics of a glow discharge. Schematic diagram of the experimental installation is shown in figure 1.

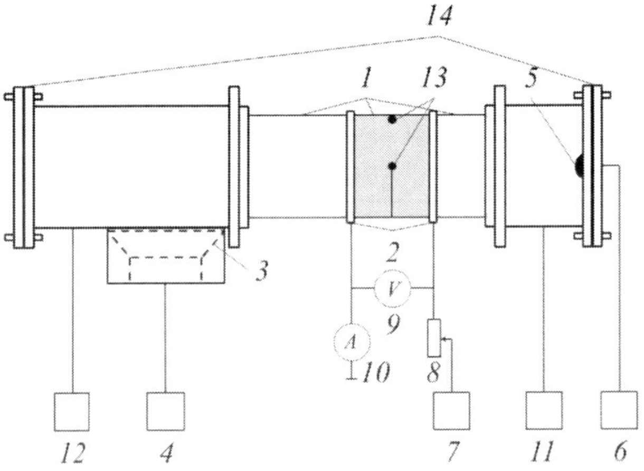

Figure 1. Schematic diagram of the experimental installation.

Experimental installation includes a set of quartz tubes 1, Copper electrodes 2 (cooled with running water), electroacoustic device 3, LFO 4, a microphone 5 (which can move along the inner wall end), Oscilloscope 6, a high voltage DC source 7, ballast resistance 8 (designed to stabilize and prevent the disruption of a glow discharge in an electric arc), voltmeter 9, ammeter 10, vacuum pump 11 and a pressure gauge 12. To measure the radial temperature gradient in the discharge gap have two thermocouples 13 on the wall and the axis of the chamber. Detachable ends of discharge chamber 14 designed for mounting the optical resonator.

Since we are using the system kilowatt power with large diameter, we have an opportunity install optical resonators of different types – the stable and unstable [13]. In the case of block with an unstable optical cavity installation is provided to output radiation. Block contains intracavity aperture mirror mounted at an angle to the optical axis 45°. Intensity distribution of output radiation in the near field has the shape of a ring. Schematic diagram of the optical unit to output radiation is shown in figure 2.

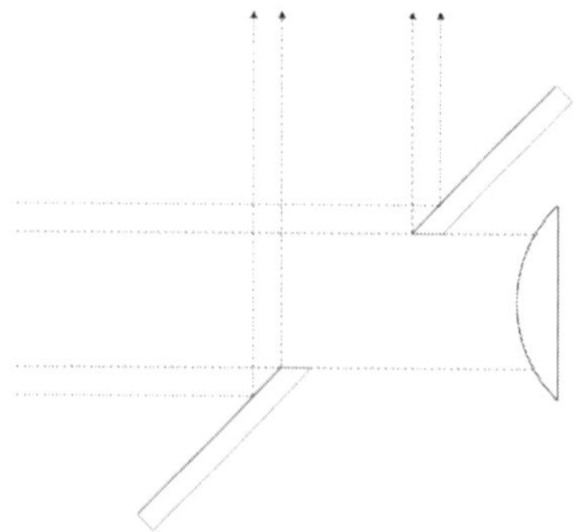

Figure 2. Schematic diagram of the optical unit to output radiation.

To achieve low vacuum sufficient mechanical pumping means [6]. In the vacuum system of the experimental installation, the pumping means is a rotary vane pump 11. There is the possibility of organizing a mean flow of gas through the discharge gap. The main parameters of the experimental installation are shown in table 1.

Table 1. The main parameters of the experimental installation

Parameter	Value
Voltage range	0-30 kV
Pressure range	1 до 760 torr
Sound intensities range	от 80 до 130 Db
Frequency range of the acoustic oscillations	от 50 до 20000 Hz
The radius of the discharge chamber	0,046 m

4. Conclusions

The geometrical dimensions of the discharge chamber allow to excite acoustic oscillations of different modes at sound frequencies is perceptible by humans. Longitudinal acoustic oscillations are the most common type of sound in nature. A standing wave of longitudinal acoustic oscillations is characterized by alternating successive maximum and minimum of the pressure and velocity along the axis of the chamber. Particle displacement occurs along the discharge chamber. Transverse acoustic modes of oscillations are of three types – the tangential, radial and combined. For these modes the displacement along the axis of no gas oscillates only in the direction perpendicular to the axis of the discharge chamber. Combined modes of transverse oscillations are mixed oscillations that are tangential component and radial component of the vibrational velocity. The frequency of these oscillations is proportional to the order transverse modes. According to the parameters of the experimental installation is given in table 1 in the discharge chamber can excite modes tangential acoustic oscillations of the twelfth order and modes radial acoustic oscillations of the fifth order. Combined modes can be realized when excited tangential and radial oscillations of the third order together.

Visualization of processes occurring in the discharge chamber, measuring the current-voltage characteristics and the temperature gradient between the axis and the wall of the discharge chamber allow an assessment of the thermal conductivity in the radial direction, the energy input into the glow discharge and the volume of plasma formations.

The work is performed according to the Russian Government Program of Competitive Growth of Kazan Federal University.

References

[1] Kovalev O B and Fomin V M 2013 *Physical fundamentals of laser cutting thick sheet materials* (Moscow: "FIZMATLIT") p 256
[2] Fadeev S A and Kashapov N F 2013 *J. Phys.: Conf. Series* **479** 012009
[3] Gavrilova V A, Fazlyyyakhmatov M G and Kashapov N F 2013 *J. Phys.: Conf. Series* **479** 012010
[4] Gavrilova V A, Kashapov N F and Kashapov R N 2011 *Biomedical Engineering* **45** 198–200
[5] Dautov G, Dautov I, Fayrushin I and Kashapov N 2013 *J. Phys.: Conf. Series* **479** 012001
[6] Raizer Y P 1992 *Gas Discharge Physics* (Moscow: "Science") p 536
[7] Kashapov N F and Israfilov Z K 1991 *Journal of Engineering Physics* **60** 364–368
[8] Larionov V M and Zaripov R G 2003 Gas oscillations in installations with combustion (Kazan: KSTU) p 227
[9] Iovleva O V and Larionov V M 2012 *Russian Aeronautics* **55** 388–391
[10] Aramyan A R and Galechyan G A 2007 *Physics-Uspekhi* **50** 1147–69
[11] Aramyan A R, Galechyan G A and Manukyan G V 2008 *Acoustical Physics* **54** 774–777
[12] Kashapov N F and Fadeev S A 2014 *Izv. VUZ. Fizika* **57** 116–119
[13] Witteman W J 1987 *The CO_2 Laser* (Berlin: Springer-Verlag) p 309
[14] Weston G F 1985 *Ultrahigh vacuum practice* (London: Butterworths & Co.) p 288

Development of the method of creation of the 3d model of topology of the surface by means of the optical microscope

R.R. Gaisin[1], A.V. Nikiforova[1], N F Kashapov[1] and R N Kashapov[1,2]
[1] Kazan Federal University, Kazan, 420008, Russia
[2] Laboratory of Radiation Physics, Kazan Physical-Technical Institute of the Kazan Scientific Center of the Russian Academy of Sciences, Kazan, 420029, Russia

E-mail: kashramiln@gmail.com

Abstract. In article the question of three-dimensional visualization of topology of a volume body on its parallel cutoffs, the dynamic processes received by means of an optical microscope for the subsequent display is considered.

1. Introduction

Rapid development of new methods of processing of a surface of materials [1] by means of plasma methods requires creation of new, simple and cheap control methods. At the moment more and more the method of plasma and electrolytic formation of a microrelief of a surface is widely adopted [2]. As the method of plasma and electrolytic processing has a large number of the various modes of processing in case of which use big variety of morphological surfaces [3] turns out, there is a need of creation of a mathematical model of dependence of the received surfaces on the processing modes.

It can be reached, having received empirical dependence of system of equations, describing a surface, from the processing mode. For receiving system of equations, describing surface morphology, it is necessary to develop a method of creation of a 3D model of topology of a surface. This technique will be based on research of a standard laboratory optical microscope of the optical camera. At the initial stage of visualization there is a focusing of a microscope at surface "peaks". Selection of the focused "even" sections and record of their location in data array happens by a program method in the image. Thus, we received the first cutoff of a surface. Further with a certain step scanning of a surface is made. As a result we receive a large number of cutoffs from which it is necessary to create a 3D model. Development of a new technique of creation of a 3D model from parallel cutoffs of a surface is the purpose of this article.

Problem definition. There is a collection of the parallel cutoffs received as a result of research of a standard laboratory optical microscope of the optical camera which is stored in the file in the form of coordinates of points of each of cutoffs. It is necessary to construct three-dimensional model on these cutoffs, using: a method of polygon grids, a set of APIYAPRANA libraries for achievement of independence of three-dimensional display of a platform.

2. Experimental

The algorithm was realized in the *ActionScript 3* language as one of conditions of an objective was use of *API YAPRANA* libraries which are the cross-platform tool for development of three-dimensional dynamic browser online applications of real time and are a packet of high level class libraries in the *ActionScript 3* language for use in Flash (ver.CS5 above) and Flex (ver.3.5 or above).

We will enter determination of some distances used in algorithm:

Determination 1. The distance from a point of one cutoff to an edge of other cutoff is calculated as the amount of distances from this point to each of the ends of the considered edge.

Determination 2. The distance from a point of one cutoff to other cutoff expresses through the amount of distances from this point to each of cutoff points.

Determination 3. The distance from one cutoff to other cutoff expresses through the amount of distances from each point of one cutoff to each of points of other cutoff.

Determination 4. The distance from an edge of one cutoff to an edge of other cutoff expresses through the amount of distances from each point of one edge to each of points of other edge.

Theoretical proofs to these statements it isn't found yet, but use of these axioms proves correctness of their assumptions in practice.

The developed algorithm has possibility of restoration of three-dimensional model of the researched object on the cutoffs located parallel to three main projectional planes accepted in medicine: sagittal (OXZ), frontal (OYZ) and axial (OXY), thanks to the appropriate matrixes of transition from "real" coordinate system to "worker" as whom the axial plane is accepted. As coordinate system of algorithm the left Cartesian rectangular coordinate system is selected.

The projectional plane is determined by a normal vector of the plane constructed on three any points of one of model cutoffs. So, for example, the normal vector of the axial Oxy plane has appearance $\bar{\boldsymbol{N}} = \mathbf{(0,0,const)}$, sagittal Oxz- $\bar{\boldsymbol{N}} = \mathbf{(0,const,0)}$, frontal Oyz- $\bar{\boldsymbol{N}} = \mathbf{(const,0,0)}$, respectively.

One of the principal advantages of algorithm is visualization of three-dimensional nonconvex and/or porous object on a collection of parallel cutoffs, to each of which there corresponds the different quantity of peaks.

The basis of the solution of the initial task was formed by the principle of coercion of cutoffs to the generalized cylinders provided in Zhizher P article. In "Methods and algorithms of creation of 3-D models on cutoffs" [4], but with some corrections. In particular, the method described in [4] assumes equal quantity of points on cutoffs, and conditions of an objective don't provide restrictions on quantity of points, and it means that the quantity of points of cutoffs can be different. Owing to adjustment of this algorithm of restoration of objects according to conditions of the initial task, there was a row of problems which were solved as follows:

Connection of two cutoffs with different quantity of points, connection of type «1:1»

We will designate a cutoff with the greatest number of points as cut1, and with the smallest, respectively, as cut2. We move the beginning of coordinate system to center of masses of a cutoff of cut1, and the beginning of coordinate system of a cutoff of cut2 - to the point which is precisely over center of masses of a cutoff of cut1.

Note: The center of masses of system of points is calculated on a formula:

$$\vec{r_c} = \frac{\sum_i \vec{r_i}\, m_i}{\sum_i m_i},$$

where m_i— masses of points, $\vec{r_i}$— their radius vectors (setting their situation concerning origin of coordinates), and $\vec{r_c}$— the required radius vector of center of masses.

At first we will consider all points of a cutoff of cut2, for each of which we will select the closest edge of a cutoff of cut1, remembering thus compliance of peaks. Then we consider the remained edges of a cutoff of cut1 and for each of them we select the next cut2 cutoff peak, adjacent to it, remembering again thus compliance of peaks. The final stage is restoration of missing edges according to the received list of compliances from which the peak of a cutoff of cut1 is selected and

IOP Conf. Series: Materials Science and Engineering **69** (2014) 012007 doi:10.1088/1757-899X/69/1/012007

cut2 cutoff peaks, adjacent to it, are looked for. Each similar three of peaks forms as a result a triangulated edge.

Ramifying implementation - connection of cutoffs of type «1:m», «m:n»

This problem is solved with the help of an additional cutoff which is located between the initial cutoffs from each of group of cutoffs equally spaced and built by paired combining of a set of cutoffs through points of their centers of masses. In the course of creation of an auxiliary cutoff the special vector which components are the copies of cutoffs participating in combining procedure is created. The appropriate centers of masses are added to each cutoff of this vector. The additional cutoff plays a role of the intermediary and c the initial cutoffs as follows connects:

with the single cutoff as connection «1:1»

for connection with group of cutoffs the special vector which each cutoff connects on the «1:1» type to the appropriate cutoff from a set of the initial is used.

Triangulation of cutoffs

The circuit of a nonconvex cutoff breaks into convex circuits, each of which then pass a triangulation which results integrate in uniform result.

The triangulation of a convex cutoff consists in processing of a vector of indexes of the peaks written as their sequential numbering of a cutoff:

If the considered vector contains odd quantity of peaks, we pass to point 2, in an opposite case - to point 3. If the circuit consists of three peaks, it is considered a triangulated edge and indexes of its peaks are simply added to a total vector of a triangulation of model.

We integrate peaks in groups on three and every time we move on 2 line items in a vector, saving thus the first peaks of each similar group in a special auxiliary vector which is processed similarly, starting with point 1 (thus the last peak of the initial vector is considered the first peak of group and also it is added to an auxiliary vector). This auxiliary vector every time is updated and processed until begins to contain <=3 components. If length of an auxiliary vector becomes equal we rub, it is entirely added to a total vector of a triangulation. The total of a triangulation are the received groups of indexes which are added to a total vector of a triangulation of model.

We add its first component to vector head and further we process it similar to point 2, except for processing of an auxiliary vector - in which before processing it is necessary to delete the last element.

Creation of porous model

For determination of a type of model according to input data it is necessary to clarify layout of cutoffs relatively each other. The principal condition of the input file is superiority of an external cutoff in relation to the remaining internal cutoffs - "holes" belonging one plane in case of porous model of the researched object. Therefore when processing the input file with a set of coordinates of peaks of cutoffs there is a processing of each similar peak based on the complex analysis to the purpose to clarify, whether it gets to a circuit of an external cutoff. If all peaks of the considered cutoff get to a circuit of an external cutoff with which they lie in one plane, this cutoff is considered internal.

For a case of porous model of the researched object connection of cutoffs happens as follows: internal cutoffs of one plane connect to internal cutoffs of other parallel plane corresponding to it, and connection can be several types: one to one, one-to-many and many to many. External cutoffs connect to the external cutoffs of the parallel plane corresponding to them. In case of implementation of similar connection of cutoffs it is necessary to observe the correct bypass of peaks of circuits for saving visibility of model in the conditions of use of *ActionScript 3* as the main programming language.

The triangulation of a porous cutoff can pass on one of the following options

If the external cutoff has one internal cutoff: Total triangulation = a triangulation of connection of the «1:1» type of an external cutoff with internal ∪ a triangulation of an internal cutoff.

If the external cutoff has more than one internal cutoff:

Total triangulation = a triangulation of connection of combining of internal cutoffs through centers of masses of system of points of their two edges closest to each other with an external cutoff ∪

IOP Conf. Series: Materials Science and Engineering **69** (2014) 012007 doi:10.1088/1757-899X/69/1/012007

a triangulation of combining of internal cutoffs through centers of masses of system of points of their two edges closest to each other.

3 Inference

Uniqueness of the developed algorithm consists in possibility of simulation of both the convex, and nonconvex researched objects, and also in creation of three-dimensional treelike model on coordinates of points of parallel cutoffs in the conditions of absence of any restrictions on quantity of peaks of each of cutoffs. The method of polygon simulation used thus is simple in implementation and allows to avoid in case of creation as shadows, and mirror flares. Thanks to it transmission quality of the form of object, and consequently also quality of researches increases. Use of API YAPRANA libraries allows to create on the basis of the developed algorithm the online instrument of three-dimensional visualization of objects on a collection of parallel cutoffs for the subsequent display of dynamic processes without restrictions on quantity of polygons, which is calculated automatically in the course of creation of a 3D model. Besides API YAPRANA libraries allow to reach platform independence of the final product and, as a result, promote its active implementation and distribution in different spheres of researches.

References

[1] Plasma application of protective polymer-powder coatings to ultrasonic sensors. BiomedicalEngineering.2011. T45(5) 138-200
[2] Research of plasma-electrolytic processes in various ratio of the area anode to cathode. Journal of Physics : Conference Series. 479(1), 012005, 2013
[3] Investigation of the influence of plasma-electrolytic processing on the surface of austenitic chromium-nickel steels.479(1), 012003,2013.
[4] Zhizher P. V. Methods and algorithms of creation of a 3D model on cutoffs.
[5] Atlas of scanning of a brain.Web: http://www.info-radiologie.ch/atlas-tdm-cerveau-ru.php#.
[6] Ivanov V.P., Batrakov A.S. Three-dimensional computer graphics . 1995. — 224 c.
[7] Visualization of volume in medicine - № 05, 1996/ Open systems / Web: http://www.osp.ru/os/1996/05/178989

Mini-Central heating and Power Plant (CHP): the choice of the optimal structure and modes of operation

I R Gil'manshin[1], I A Konahina[1], N F Kashapov[1] and N N Fahreev[2]
[1]Kazan Federal University,
420008, Kremlevskaja Street, 18, Kazan, Russia
[2]Kazan State Power Engineering University,
420066, Krasnosel'skaja Street, 56, Kazan, Russia

E-mail: chelny@kpfu.ru

Abstract. The paper provides an overview of the different types of organizations of mini-CHP, the results of a comparative evaluation of their performance in terms of the variable load. The recommendations to improve power facilities of the mini-CHP on the basis of system analysis methods and the technical and economic optimization are defined.

1. Introduction

Boiler systems are the most common sources of thermal energy in Russia. Large district heating plants with an installed capacity of 100 MW supply with warmth hundreds of apartment buildings and an important thermal power facilities in the infrastructure of cities. In Kazan, for example, 122 district boiler houses operate.

2.Main part

Power supply (in this case electricity) of boilers themselves for their own use (pump drives, blast equipment, cooling fans) is realized from an external source, which is often the only way for the boiler house. Power failure in the boiler room in the winter period and the subsequent shutdown of equipment can lead to consequences connected with heavy losses. Consequently, the most promising solution to ensure the reliability and efficiency of boiler plants, including covering their own needs in electricity and additional sales of surplus power generated in the overall network is setup on existing boilers generators, drive, which can be a gas piston or gas-turbine engine (GTE) with hardware of flue exhaust gas utilization (mini-CHP). Organization of mini-CHP based on district boiler increases the reliability of power supply, thus energy loss during transportation and the cost of fuel for production of heat and electricity are reduced to a minimum. Building of structure of mini-CHP can be varied. Image 1 shows the possible options of CHP structure applied to the reconstructed and newly projected sources.

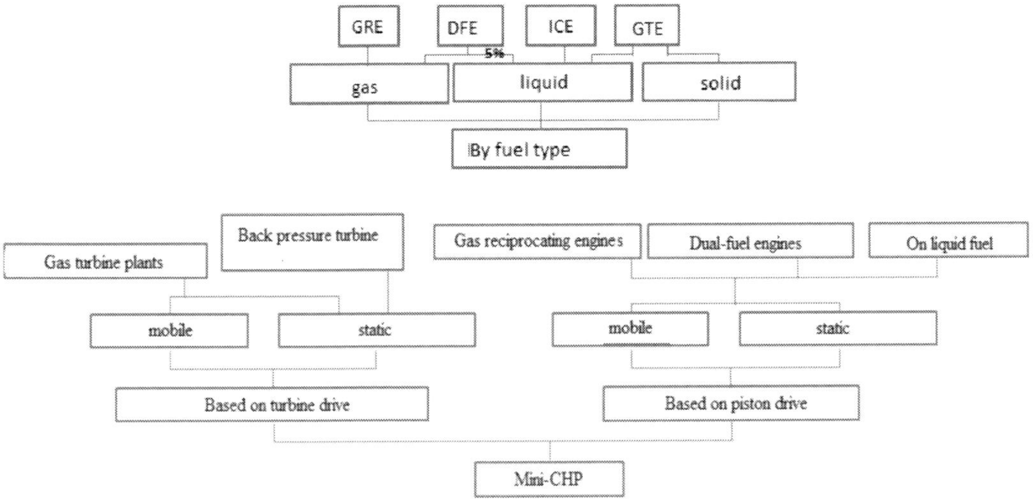

Im. 1. Variants of organization of mini-CHP.
GRE- gas reciprocating engine; DFE- dual-fuel engine;
ICE- internal-combustion engine; GTE- gas turbine engine.

As seen in Image 1, for fixed installations, gas-fueled, competing options are mini-CHP based on GTE and GRE.

Heat boiler schedule is characterized with big unbalance. This leads to difficulties in selecting the operation mode of the cogeneration superstructure. It is known that the greater the number of hours electricity is generated on the basis of caloric intake per year, the higher the efficiency of the cogeneration settings (KoGS). Year-round thermal load in the boiler houses is only hot water (DHW). In this connection, the thermal power cogeneration unit of mini-CHP is advisable to be choosen based on the conditions covering the DHW load in the summer period, which provides its year-round operation and, therefore, the most effective use. On the other hand, the relative capital costs of establishment of power generation set decrease with the increasing of their unit power. Therefore, for the reconstruction of boilers in mini-CHP firstly it's necessary to choose the largest of them [2,3].

Similar object is a thermal boiler of Kazan city, located at Kashtanovaja Street, 18 (2a). Boiler room is equipped with boilers brand KVGM-10 at the number of 1 unit and HRG-8 at 3, with the pump: ZN-400-105 – 3 units, D-315-71- 3 units. Connected heat load of heating is 21.96 Gcal / h, of ventilation is 0.34 Gcal / h, DHW 3.58is Gcal / h.

This boiler was selected as the base object in research. Due to the fact that the boiler operates on natural gas, as options of its restructuring reorganization was considered superstructure based on GTE on the type GTES- 25P and based on gas engines brand of YMZ -8503 (Image 2b).

IOP Conf. Series: Materials Science and Engineering **69** (2014) 012008 doi:10.1088/1757-899X/69/1/012008

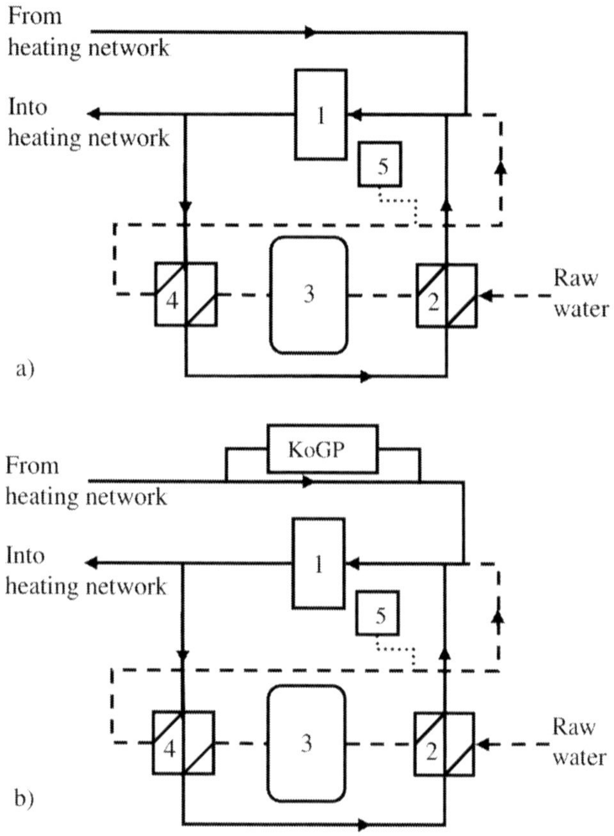

Im. 2. The scheme of the original water boiler and synthesized on the basis of its mini-CHP
a- the original scheme; b- synthesized mini-CHP with cogeneration units:
1- water-heating copper; 2- the raw water heater; 3- chemical clearing; 4- heater of the chemical cleared water 5- reagent tank

Optimization of regime parameters of mini-CHP and its structural organization was conducted based on the methodology of System Studies of industrial CHP, which has been applied to the new object - water boiler house, translatable in mini-CHP mode [4, 5]. The method includes the analysis of the efficiency of heat and power facility, structural modeling of the synthesized object with improved performance and optimization of mode parameters on the selected criterion of efficiency.

In this case a power-economic optimization was held. The extremum (maximum) function was the object of search

where – reached fuel saves compared to the base case (heating from the boiler house, power supplying of boiler house from condensing power plant CPP) - the coefficient of the synthesized cogeneration of mini-CHP.

Optimization calculations were carried out by approved methods described in [5].

3.Conclusion

Comparative analysis of the results showed that the best performance is the structure of mini-CHP with GRE. Compared with the alternative base variant optimized mode provides 46% fuel savings. Option with GTE was less effective, but compared to the base case it also gave 41% fuel savings.

When working optimally mini-CHP based water boiler house allows you to fully meet its own needs for electricity in summer period and 50% of the needs of the boiler during the heating period.

4.References

[1] Kotler V R 2006 *Mini-CHP: foreign experience*.- Thermal Engineering, №8
[2] Fahreev N N Shakirova L R 2013 *Superstructure with cogeneration units of water boilers in conditions of the modernization of district heating systems*.- Abstract in 4 parts. P. 3. - M: Publishing House of MEI
[3] Direktor L B Popel' O S Frid S E 2010 *Analysis of the effectiveness of mini CHP-based ICE at variable heat loads*.- Electronic Journal of Energy Service Company "Environmental Systems", №3 (99)
[4] Melent'ev L A 1983 *System studies in engineering. Elements of the theory, development trends*. 2nd Edition - M: Nauka
[5] Nazmeev J G Konahina I A 2002 *Thermal energy systems and energy balance of the industrial enterprises*. - M: Publishing House of MEI

IOP Conf. Series: Materials Science and Engineering **69** (2014) 012009 doi:10.1088/1757-899X/69/1/012009

Heightening of efficiency fuel using installations due to external recuperation units of waste heat

I A Konahina[1], I R Gil'manshin[1], T R Safin[2]
[1]Kazan Federal University,
420008, Kremlevskaja Street, 18, Kazan, Russia
[2]"Gazprom mining Nadym"
629730, Nadym Zvereva st., 1 Yamal-Nenets Autonomous Region, Russia

Abstract. Energy saving activity on installation of heat exchangers for utilization of afterheat of smoke fumes of industrial furnaces and steam generating units is offered. Problems of their maintenance and methods of elimination of the given problems are considered.

1. Introduction

The main flow of heat in industrial furnaces and boiler installations occurs during fuel combustion. However, if taken the amount of heat received by the burning of fossil fuels as 100%, then directly to the conduction of the technological process in the furnaces and the subsequent heating of the water in the heat recovery boiler (installation of them on the flue gas stoves is mandatory) and for heat generation in boiler units, is consumed only 70 to 85%. The remaining amount of heat energy from 15 to 30% is lost. Efficiency of steam generating units is determined by the formula,%:

$$\eta_{\text{т}} = 100 - (q_{\text{ух}} + q_{\text{н.о.}} + q_{\text{х.н.}}) \tag{1}$$

where $q_{\text{ух}}$ - losses from the outgoing flue gases, $q_{\text{н.о.}}$ - losses of external cooling structures; $q_{\text{х.н.}}$ - losses due to incomplete chemical combustion of fuels;

As shown in [1], the most important role (about 75% in the total heat loss) play the losses with the exhaust gases, which is approximately determined by the ratio,%

$$q_{\text{ух}} = \frac{t_{\text{ух}} - t_{\text{в}}}{\theta}\left[C + (h-1)\cdot K \cdot B\right]\cdot 100\%, \tag{2}$$

where $t_{\text{ух}}$ – temperature of the flue gases exiting the boiler, °C; $t_{\text{в}}$ – temperature of air entering the boiler, °C; θ - theoretical adiabatic combustion temperature without thermal effect of dissociation of the combustion products, °C; C – correction coefficient of the specific heat of combustion products; K – correction coefficient of the heat capacity of air; B – a correction factor for the volume ratio of dry and wet combustion products from the combustion of fuel in a stoichiometric amount of air; h – the ratio of the theoretical volume of the dry gas to the actual volume.

2. Basic part

Mainly q_{yx} depends oh temperature $t_{yx,}$ however, the cooling gas in conventional systems below 120-180 C is not realized because of the exploitation nature of problems:

1. Due to the condensation of water vapor and fuel oxidation products which enter into a chemical reaction with the formation of acid on heat transfer surfaces of boilers and ducts, we can observe an acid corrosion;
2. The cooling gases reduces the driving force of the flue gases to the atmosphere through the chimney.

   As an external heat recovery boilers in [2, 3] are considered the contact type devices, however, and their use is associated with a number of negative factors that should be considered when selecting the event:

1. Cooling of gas below the dew point temperature is not economically feasible;
2. Transfer processes in heat exchanger "ballast" (reverse) heat are not excluded;
3. Water in contact with gas is saturated by carbon dioxide which increases its corrosiveness;
4. Installation of the contact heat exchanger creates a significant hydraulic resistance, the overcoming of which requires the additional energy.

Along with the negative factors, there are some positive:

1. Common unit of heat recovery is created, which allows to conduct process automatically in conditions that ensure the highest technically achievable efficiency and safety of the process;
2. Condensation of water vapor in the heat exchanger-utilizer reduces the viscosity of the flue gases, thus reducing energy costs for transporting them through the gas duct. As it is shown in [3], in some cases, the installation of the heat exchanger to exhauster can improve the hydraulic characteristics of the gas path, even in comparison with the original version (without heat exchanger)

Taking into account the above factors the recycling unit scheme is designed based on surface heat exchanger-utilizer of condensing type (Fig. 1). The flue gases are cooled in a surface heat exchanger to a temperature of 40-45 C, heating the feed water to an estimated temperature (raw-water or chemically treated water can be heated).

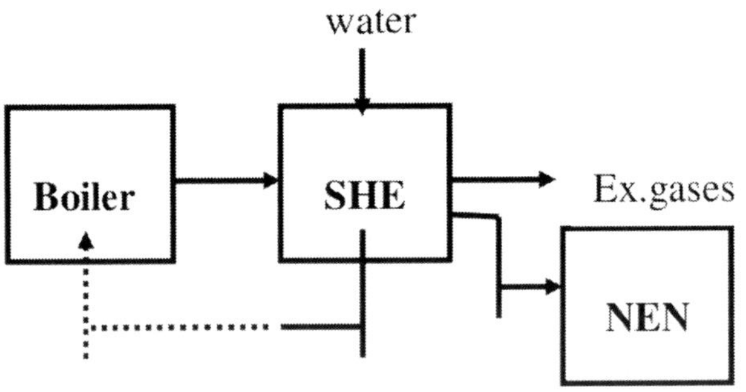

Fig. 1. The heat recovery unit
SHE - surface heat exchanger; NEN - node of effluent neutralization

Condensate is piped to the node of neutralization, where processes of the concentration of effluents and their preparation for recycling are conducted.

As the base object for setting the recycling node is selected boiler of "Gazprom mining Nadym" Ltd. With hot water boilers KBM-5.2 MW, which is supplied with natural gas of Cenomanian mine of Yamsoveyskoye field. Conducted analytical calculations have shown that the proposed system allows to improve the thermal efficiency of the boiler by 8.5%.

3.References

[1]. Nazmeev YG, Konakhina IA *Thermal energy systems and energy balance of the industrial enterprise 2002* Moscow: Publishing MEI.

[2]. Galustov VS *Heat and Mass Transfer processes and devices with direct phases contact in Heat and Power Engineering* 2003 Energy and management № 4.

[3]. Grigorov V G, Neyman V K, Churakov S D *Utilization of low-grade thermal waste energy in chemical plants.*1987 M .: Chemistry.

IOP Conf. Series: Materials Science and Engineering **69** (2014) 012010 doi:10.1088/1757-899X/69/1/012010

Energy service contracts in regional engineering center for small and medium businesses

I R Gil'manshin, N F Kashapov
Kazan Federal University,
420008, Kremlevskaja Street, 18, Kazan, Russia

Abstract. The analysis of the energy service contracts development in Russia is given in the article. The role of the Complex learning centres in the field of energy efficiency in the promotion of energy service contracts is described. The reasons of constraining the development of energy service contracts are described.

1. Introduction

Energy conservation and energy efficiency are the key ingredients to ensure competitiveness of the modern economy. In order to create the legal, economic and organizational basis to stimulate energy conservation and energy efficiency President of Russian Federation D. Medvedev signed a law FL-261 "On energy saving and energy efficiency improvements and on Amendments to Certain Legislative Acts of the Russian Federation."

In the FL-261 the main activities to ensure energy efficiency are determined: the information base formation, analysis and evaluation of energy efficiency, identification of ways to improve energy efficiency, the development of energy efficiency program, the introduction of advanced technologies and solutions in the field of energy conservation and outreach services to citizens.

2. Basic part.

Effective implementation of the requirements of the law requires:

- The presence of a sufficient number of qualified professionals with expertise in the field of energy efficiency and skills of practical integration of advanced technologies and solutions to existing engineering networks,
- Rooting in the mass consciousness of citizens the principles of energy efficient behavior,
- Construction and testing of a system of external funding for energy efficiency activities,
- Organizing information platform of interaction of consumers and suppliers of energy efficient products.

To solve these problems in Kazan (Volga) Federal University and 8 educational institutions in Russia, with the support of the Russian Ministry are created the comprehensive learning centers in the field of energy efficiency (hereinafter - the Center). [1]

A wide range of tasks are solved in the created centers.

1. Implementation of additional educational programs and educational services in the areas of energy efficiency, energy conservation, energy efficiency and environmental management.
2. Carring out the energy audits.
3. Carrying out a research work in the field of energy saving, energy efficiency and environmental management.

4. . Implementation of organizational performance and management of the system for energy efficiency in the KFU.
5. Development and testing of methods of effective interaction of organizations in the field of energy audits, engineering companies, investors and customers in order to promote energy service contracts.
6. Popularization of energy saving and energy efficiency, the creation of the presentation site of advanced technologies and solutions in the field of energy conservation and energy efficiency.

An important aspect of the implementation of FL-261 is to provide energy efficiency standards of energy saving measures in the design, implementation and analysis of the results [2]. In the absence of a systematic approach to the implementation of measures to improve energy efficiency the achievement of targets, in some cases is unattainable. In many respects by the lack of qualified personnel at the customer [3,4] and effective, transparent implementation of the methodology for energy efficiency is explained the restrained development of energy service contracts in Russia [5,6]. Investors are forced to conduct a comprehensive examination of the project or organize a full cycle of analysis and design work [7], which significantly increases the implementation time and reduces the attractiveness of the project. Development of a flexible mechanism for the formation of verification and maintenance of energy service contracts will significantly shorten the period of preparation of documentation, reduce indirect costs, ensure the transparency of the project.

However, the great importance of the implementation of energy efficiency measures has the educational and outreach work [1]. Education and awareness-raising activities are held in the form of training seminars or training courses. During which the legal basis of energy saving, the basic theoretical issues, specific examples of energy-efficient technologies and solutions are disclosed. Maintaining the relevance of acquired knowledge is achieved by placing visual-campaign materials.

One of the main factors hindering the modernization of energy equipment and increasing its energy efficiency is the difficulty in attracting foreign investors. Investor, in turn, for accepting a positive decision needs the researched, transparent project with verification. Implementation of these requirements is possible with the implementation of measures to improve energy efficiency through energy service contracts of a full cycle (Fig. 1). Moreover, the development and implementation of energy service contracts in the framework of a regional engineering center significantly increases the attractiveness of the project to the investor.

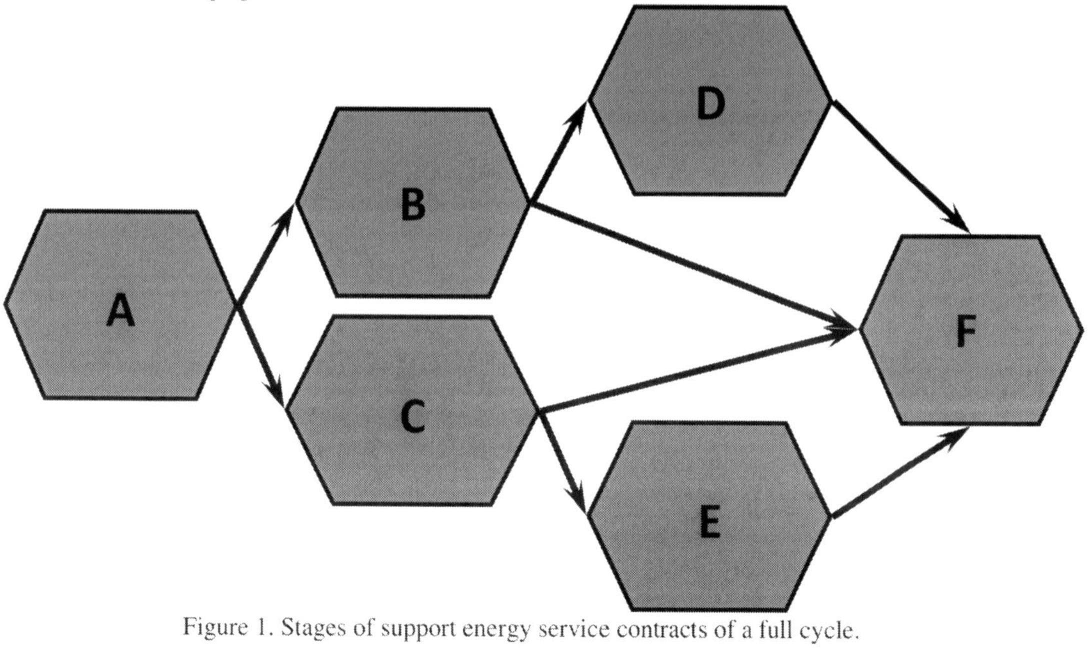

Figure 1. Stages of support energy service contracts of a full cycle.

A - formulation of the problem, B - consulting. C - project development. D - project funding. E - maintenance / technical supervision. F: Supply of equipment / installation.

Moreover the created in a regional engineering center target area of interaction of highly qualified specialists, companies implementers, developers and manufacturers of energy efficient products will effectively solve complex, multifaceted problem. Complex training centers in the field of energy efficiency in this case, will assume the role of information consolidators of information to ensure coordination of effort and resources to solve a problem within the energy service contracts. The scheme of interaction of the customer and the Centre is reflected in Fig. 2.

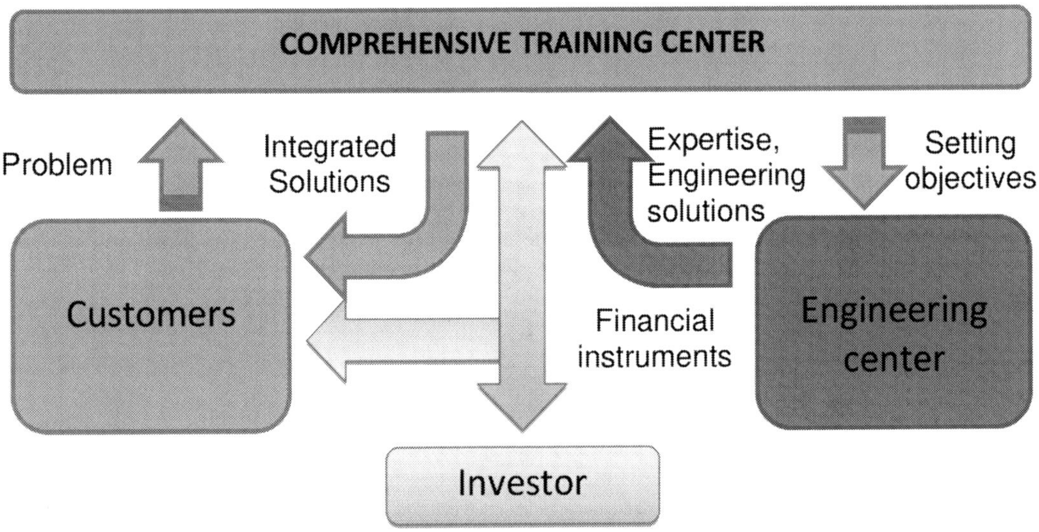

Fig. 2. Scheme of the interaction of the customer and the Centre.

In accordance with the proposed scheme a customer formulates a unified interaction task to achieve a certain result. Further, the Center made its decomposition on common tasks and select appropriate solutions. Simultaneously, the collection and accumulation of the necessary general and statistical information is carried out. After passing the preliminary examination a search for an investor is conducted. By this time, the Center has already collected the necessary amount of data objectively reflecting the efficiency of the project. There is the possibility to obtain current data in remote access regime. The collection of this information will allow a potential investor in a short time to make a decision on the financing of the project. Further, the project moves into the implementation phase, the choice of equipment, preparation of project documentation, direct mounting. At the final stage, the technical supervision and analysis of the degree of achievement of planned targets are carried out, and if it is necessary the remedial works are implemented.

The proposed methodology to support energy service contracts and the customer and Centre interaction scheme are providing the required efficiency and transparency of the process of implementation of energy efficiency measures, which will certainly contribute to the popularization of energy service contracts.

3.References

[1]. *The network of educational centers for energy conservation and energy efficiency* [electronic resource]. - Mode of access: http://energycouncil.ru/.

[2]. Koval S. *Energy saving in the enterprise. The introduction of the standard ISO50001* [electronic resource]. - Mode of access: http://portal-energo.ru/articles/details/id/136.

[3]. Tatarnikov AV *Energy activities in the implementation of the "State Program" Energy in Moscow 2011, 2012-2016. and up to 2020.* 2012 "// Information resources of Russia. №1. S. 26-28.
[4]. Ivanov *Energy service contracts. Realizing the potential of energy saving* [electronic resource]. - Mode of access: http://portal-energo.ru/articles/details/id/378.
[5]. Tsakunov C *Implementation of energy service contracts in Russia* [electronic resource]. - Mode of access: http://www.abok.ru/for_spec/articles.php?nid=5231.
[6]. Antonovich S *Energoservice: Problems and positive examples.* - Mode of access: http://udee.energohelp.com/articles/energo_services/5274/.
[7]. Usievich V *Heating companies and energy services for consumers: Experiences and Perspectives* Mode of access: http://www.energosovet.ru/bul_stat.php?idd=349.

Noise absorbing composite materials applied in domestic trucks

I F Gumerov[1], L N Shafigullin[1], S M Vakhitova[2], I F Shaekhova[1]
[1]*Kazan Federal University,*
423800, 68/19 Mira avenue, Naberezhnye Chelny, Russia
[2]*Scientific and technical centre of KAMAZ Inc.*
423800 Jalil avenue, 29 Naberezhnye Chelny, Russia

E-mail: chelny@kpfu.ru

Abstract. One of the basic indicators of the modern automobile is the low noise level. Noise level decrease is reached due to: 1) sources of noise elimination due to change of a design of elements and automobile systems; 2) application of modern noise insulation and noise absorption materials. The following noise absorption materials in domestic trucks are applied: fiberglass plastic, basaltic fireproof roll material (BFRM), AA SMT, AL-aralamino, isomat.

1.Introduction

Modern automobile represents the complex technical object consisting of many subsystems to which inconsistent demands are often made. Salon or a cabin noise level is an important consumer indicator of the automobile. Manufacturers of automobiles direct considerable amount of money and manpower on improvement of salon noise insulation and elimination of noise sources.

Drivers of buses and trucks are mostly subjected to influence of harmful sound effects as their intensity several times exceeds rate. Maximal possible admissible noise level in Russia is considered 74-75 dB (at frequency of 1000 Hz). In France, for trucks the admissible limit is 90 дБ [1; 2].

Loud sound effects substantially influence fatigue of the driver (organs of hearing, nervous system) that can lead to reaction and attentiveness deterioration. Long stay in a truck cabin in which negative noise constantly act on a person , can lead to increase in the latent period of impellent reaction, decrease in stability of clear vision, easing of twilight sight, infringement of concentration of attention, coordination of movement and function of a vestibular mechanism, fast fatigue. As a result, the risk of occurrence of road accident increases several times.

2.Noise absorbing composite materials

The basic sources of noise in a truck salon are: the working engine causing vibration, transmitted through elements of fastenings on motor transport body; elements of torpedo parts at their poor-quality assembly, poor-quality plastic in the truck design, badly fixed things in salon; sliding of tyres leading to a noise, getting into salon through apertures and cracks in a body, or through the poor-quality (or damaged) glass cover.

For a noise insulation of any mean of transport special screens are used, representing designs, consisting of fiberglass plastic frame and noise absorption layer on an adhesive basis, metal elements of fastening.

The technology of a noise insulation of trucks includes application of sound - and vibro-absorbing materials, and also protective casings. These materials are divided in two groups: vibro-dampers, working as weighting compounds, and struggling with vibration of body material; sound-insulators, having low heat conductivity and weak sound conductivity. The given materials perfectly absorb a sound, and also isolate a cabin of the truck from noise which arises in a motor compartment. First, vibro-dampers are fixed, and then sound absorbers are installed [3, 4, 5].

Noise insulation materials are applied at processing of wheel arches, a cabin floor, and its partition with a motor compartment. Nonflammable materials for sound insulation are used under the hood of the truck, having a special heat-reflecting layer, minimizing risk of ignition.

Noise insulators are materials having closed cell structure, allowing absorbing noise in a wide range of sound frequencies. The most widespread noise insulation material is bitoplast, being also a good thermal insulator. Some materials possess ability to push away a sound - isolon or splan [6].

Visomat, vibroplast, bimast, etc. are refered to vibro-damper group.

Noise insulation quality of the truck depends on quality of applied materials. All materials which are applied in a truck noise insulation, should meet certain requirements: at a noise insulation of under hood space materials are required which have increased resistance to hostile environment (lubricating oil, fuel, other chemical substances applied in the truck); adhesive ability of a material (influences on reliability of gluing of a composite to a surface of a part); dampers and noise insulators should possess low water absorption; small relative density of vibro-dampering materials; safety of applied materials - should not emit harmful evaporations and bring harm to health of the person; low inflammability of the materials applied to a noise insulation of engine compartments and a subsystem of an electric power supply; high degree of damping is a factor of losses within 0,001÷0,01 [7].

KAMAZ truck was used as a considered motor vehicle in which the following kinds of noise insulation screen are used: the screen of the top part of the engine; the screen of the bottom part of the engine; the screen of a gearbox.

The following noise insulation composite materials are applied: fiberglass plastic, basaltic fireproof roll material (BFRM), AA SMT, AL-aralamino, isomat [8; 9; 10; 11].

The basic characteristics of considered materials are presented in table 1.

Table 1. Materials applied for manufacturing of KAMAZ truck screens

Indicator	Fiberglass plastic NPmT TC 6-48-00204961-8-89	MBFR-C-5, MBFR-5F TC 5769-003-48588528-00	AL-aralamino PN 043-72-84	Isomat PN 027-72-88	AA 12,5 SMT TC 2292-001-42530002-00	AA 25 SMT TC 2292-001-42530002-00
Thickness, mm	3,0	5,0 (3 layers are used)	15,0	25,0	12,5	25,0
Density, kg/m^3	1750	125- 200	Unit of the area weight 2900 g*m^{-2}±10%	Unit of the area weight 2800 g*m^{-2}±10%	Weight 1 m^2, kg 0,61±0,15	Macca 1 m^2, kg 0,97±0,15

Heat conductivity Watt/(m*K) at 25±5 °C 125±5 °C 300±5 °C Coefficient, Watt*m^{-1}*K^{-1}	Heat conductivity at 20 °C, Watt/m °C 0,75 0,300-0,500	0,045 0,065 0,090	0,038	0,041	Specific heat conductivity 0,036 Has low factor of heat conductivity	Has low factor of heat conductivity
Combustibility ignition temperature, °C Self-ignition temperature, °C	Almost nonflammable according to GOST 12.1.044-89 >500	Nonflammable	It is possible to use without protection up to 100 °C, which corresponds to A-degree of combustibility-nonflammable	290-300 390-400	according to GOST 25076 (self-extinctive)	according to GOST 25076 (self-extinctive)

Extension of table 1

Sound absorption factors at frequency, 800 1000 Hz	0,013 0,014	0,3-0,97 0,5-0,90	0,72 0,64	0,65 0,75	0,41 0,55	0,56 0,74
Resistance to water, crude oil, oils, cleaners	Resistance to influence of chemically hostile environments, salt solutions	Chemical resistance to oils, acids, alkalis	Resistant	Having high chemical resistance	—	—

Note	Metalwork finishing in a place at installation of screens is accompanied by emission of a glass dust	There are no glutinous and reflecting layers, necessity of a thickness set sewing together of several layers of a material	Cutout and assembly is accompanied by emission of a glass dust	Has low *t* ignition and self-ignition	Has low *t* ignition and self-ignition	Has low *t* ignition and self-ignition

According to the data presented in table 1, it comes that applied materials do not correspond fully to a complex of the established standards and requirements, demanded to noise insulation materials applied on KAMAZ trucks (an interval of working temperatures from -70 C° to +1000 C°; heat conductivity factor - 0,37-0,41; sound absorption factor - 0,65-1,01) [12]. In this connection there is a severe need in working out of new composite materials which can be used as noise absorption materials for KAMAZ truck screens.

3.References

[1] Ilyashuk Y M 1958 *Stiffening effect of walling on the noise-isolation* /Sb. Noise control and noise effects on the body. MY 2 Lyoto, L P 57 -76
[2] *Building Regulations . Protection against noise* : SNIP II- 12 -77 : Utv.Gos.kom . USSR Council of Ministers 6/14/77 : Term introduced. effect 01.07.78/Gosstroy Rossii. - Izd.ofits. -M . SUE LAC , 1998
[3] Klyachko L N 1965 *Questions of industrial noise reduction methods soundproofing*. Author. dis. Candidate . tehn. Sciences: Chelyabinsk . Uralniistrompro draft p 24
[4] Galyautdinov R T, Kashapov N F and Luchkin G S 2002 Avtomobil'naya Promyshlennost **3** 21-23
[5] Galyautdinov R T, Kasparov N F, Luchkin G S 2003 Inzhenerno-Fizicheskii Zhurnal **75 (5)** 170-173
[6] Noise Protection: snip 23 -03- 2003. - St. Petersburg: DEAN, 2004 457 p
[7] Kravchun bp 2002 Generation and methods to reduce noise and sound vibrations Moscow: Moscow State University, 182 p
[8] http://uralbask.ru/content/view/26/48
[9] http://kompozit-sib.ru/stsplast.html
[10] http://technicalconsulting.ru/Product.aspx?Action=Detail&GoodsID=21
[11] http://www.izolex-ba.sk/al-aralamino.php
[12] Contractual job number 10 from November 20, 2012 FBGEE HPE INECA with "NPO " ROSTAR " " Development of new noise insulation screens of "KAMAZ".

IOP Conf. Series: Materials Science and Engineering **69** (2014) 012012 doi:10.1088/1757-899X/69/1/012012

Phased array based ultrasound scanning system development

R K Sagdiev[1], E S Denisov[1], Yu K Evdokimov[1], M G Fazlyyyakhmatov[2] and N F Kashapov[2]
[1] Kazan National Research Technical University named after A.N. Tupolev, 31 K. Marksa str., Kazan 420111, Russian Federation
[2] Institute of Physics, Kazan Federal University, 18 Kremlyovskaya str., Kazan 420008, Russian Federation

Email: mfazlyjy@kpfu.ru

Abstract. Multichannel ultrasound scanning system based on phased arrays development is presented in this paper. Substantiation of system parameters is presented. The description of block diagram and hardware development is presented. The combination of the self-developed receiving and a transmitting units and commercially available FPGA unit and Personal Computer can solve our scientific goals, while providing a relatively low device cost.

1. Introduction

One of the main directions of development modern systems for ultrasound scanning, diagnostics and nondestructive testing is multichannel systems based on phased arrays.

The first ultrasound scanning systems had used one-element transducer that creates an expanding acoustic field. Ultrasound beam propagates along an acoustic axis with small expansion that allows to detect objects placed angularly relative to the beam and to measure its sizes.

The main feature of the ultrasonic phased array (PA) technology is computer controlled amplitudes and phases of the piezoelectric elements excitation pulses in multi-element transducers. Piezoelectric elements can be excited to make it possible to control ultrasonic beam parameters, e.g., radiation angle, focal distance, focal spot size by means of computer program [1]. It allows to detect arbitrary placed objects and to explore whole sections of the specimen without mechanical movements.

Our research team carries out studies in the field of development new ultrasonic phased arrays, development protective matching polymer powder coating of piezoelectric elements and echo-signal processing methods [1-7]. During our researches we face a number of problem with usage commercial PA devices. Specifically most non-destructive testing instruments are expensive and, therefore, have limited field of applications. To solve a great number of scientific and technical problems, full control of PA excitation signal parameters and full access to the received echo-signals should be provided. Existent commercial devices do not provide these features in full. Thus it is required to develop sufficiently inexpensive device for ultrasound scanning connectable with computer that meets the requirement of our researches.

2. Substantiation of system parameters

The most important parameters of PA based ultrasonic scanning system are focal distance variation range, scanning angle range, and resolutions in longitudinal and transverse directions.

Minimal measurable distance is defined by duration of the sounding pulse τ, and can be estimated as:

$$d_{\min} = v\tau\,, \tag{1}$$

where v is sound velocity in the medium.

In this case duration τ for minimal distance d_{min} = 1 mm (t = 20^0 C) for austenitic steel is about 0.19 µs, for human tissue is about 0.65 µs and about 0,67µs for water.

Maximum measurement distance without taking into account attenuation in the medium is defined by the repetition time T of ultrasound probing pulses and can be estimated using the following equation:

$$d_{\max} = vT\,. \tag{2}$$

For maximum distance of 500 mm, the repetition time should be no less than 85 µs for austenitic steel and 33µs for human tissue and water.

Axial resolution (δd_z) in the direction of the ultrasonic wave propagation can be estimated as [8]:

$$\delta d_z = v\Delta\tau/2\,, \tag{3}$$

where Δτ is sounding pulse rise time. Modern hardware components (see for example LM96550 [9]) allow to achieve rise time Δτ < 3 ns. While the sampling period of ADC required for this application (typical resonance frequencies of ultrasonic sensors using for nondestructive testing is about several MHz) is greater by an order of magnitude. Therefore the longitudinal resolution is defined by sampling period of ADC. For example, if sampling period is 30 ns the longitudinal resolution is δd_z = 0,1mm for steel, and δd_z = 0,03mm for human tissue and water.

Lateral resolution in the transverse (parallel to the active aperture of the transducer direction is estimated according to the following equation [8]:

$$\delta d_x = \frac{0.11v}{f\sin(\frac{\theta}{2})}\,, \tag{4}$$

where f is a frequency of ultrasonic oscillation, θ is an angle of the measurement point position in relation to active aperture. Taking into account that:

$$\sin(\frac{\theta}{2}) \geq \frac{(\frac{L}{2})\sin\gamma}{\frac{L}{2}+d}\,, \tag{5}$$

where L is a width of the active aperture of the transducer, d is a distance between center of the transducer and focal point, γ is an angle between longitudinal axis of the phased array and a vector directed to the focal point from the center of radiating phased array. The lateral resolution for different resonance frequencies is presented in Table 1 in case when the focal point is placed on the axis being perpendicular to the array surface and passing through its center, and the transducer active aperture has width of 45 mm.

Table 1. Lateral resolution linear phased array for different medium.

The operating frequency of the piezoelement, MHz	Lateral resolution linear phased array		
	water, mm	human tissue, mm	austenitic steel, mm
2	0.226	0.235	0.9
2.5	0.181	0.188	0.72
3.5	0.129	0.134	0.51
5	0.090	0.094	0.36
10	0.045	0.047	0.18

3. Block diagram

During block diagram developing the current state of electronic components should be taken into account. At present, ultrasound scanning systems are built on large-scale integrated circuits. Each chip implements an entire multichannel hardware unit. The block diagram of ultrasound scanning system is shown in Figure 1.

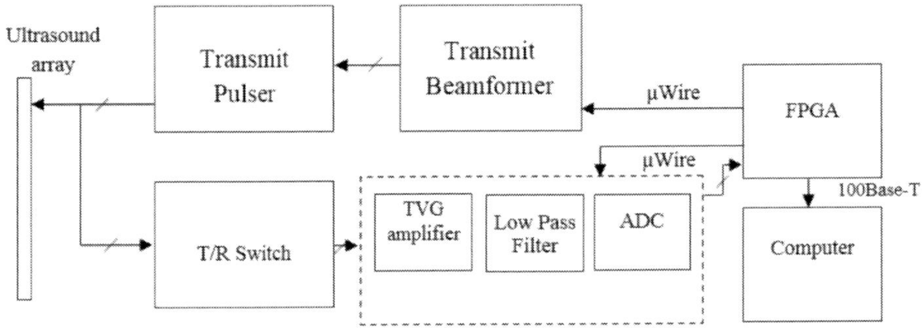

Figure 1. Block diagram of the ultrasound scanning system.

The system operates under control of computer and FPGA (Field Programmable Gate Array which is one of the most efficient programmable logic integrated circuit). Communication between computer and FPGA is implemented by means of network protocol 100Base-T or USB.

The transmitting part of the block diagram consists of a transmit beamformer and transmit pulser. The transmit beamformer generates pulses with delays individually controlled for each channel to produce a probing beam turned to the required direction or focused in the required point. The transmit pulser applies 50V voltage pulses with a specified durations and delays to ultrasound transducers. The transmit pulser is controlled by the transmit beamformer signals.

The ultrasound transducer converts electromagnetic energy of the pulses into mechanical energy of the ultrasound waves. The ultrasound wave frequency is equal to the ultrasound element resonant frequency. Thus, the frequency of the ultrasound signal is determined by the probing ultrasound elements and it does not depend on the drive pulse parameters. Reflected from the inhomogeneities in the material under test ultrasound wave returns to the ultrasound transducer. The ultrasound transducer converts received ultrasound wave into electromagnetic energy.

The return signal formed by the ultrasound transducer is provided to the transmit/receive switches (T/R Switches). It should be noted that the transducer is a device which operates with both high voltage and low voltage signals, therefore T/R Switches are necessary for the high-voltage pulse protection of highly sensitive amplifying and digitizing circuits. T/R Switches work as voltage limiter.

It keeps amplitudes of probe pulses into the permissible range up to 0.5 V. The echo signals have low amplitudes and it passes through T/R Switches without restriction.

Then the echo signal is entered into the processing unit. Processing unit comprises amplifier with time variable gain (TVG amplifier), low pass filter and analog-to-digital converter (ADC). The TVG amplifier and low pass filter are required to amplify the signal up to the level ensuring high efficiency of ADC, and to filter noises and signals above Nyquist frequency. The time variable gain is required for additional amplification of the echo signals reflected from farther objects. These signals came later and they are weaker than the signal reflected from closer objects.

FPGA controls the parameters of all units (except the computer). Furthermore, the FPGA receives a data stream from a high-speed ADC, performs primary processing of the received data and transmits the data to the computer. The computer controls operation of the entire system, performs processing of the received data and displays the results.

4. Hardware components selection

Having selected hardware components the following has been taken into account:
- Compliance of chip parameters to the aimed parameter of the ultrasound device;
- Sufficient simplicity of chip mounting on printed circuit board (PCB);
- Chip availability on the local market and acceptable delivery time.

AFE5851 (Texas instruments) [10] was selected to implement TVG amplifier, low pass filter and ADC by means of one chip. The ADC outputs are serialized in LVDS (low-voltage differential signaling) streams. In Table 2 is shown its main parameters.

Table 2. Main Parameters of AFE5851

Parameter	Value
Number of channels	16
Variable Gain	–5dB to 31dB
Input Noise	5.5 nV/√Hz
Low Pass Filter	3rd Order Anti-Aliasing Filter With Programmable Cut-Off Frequency (7.5, 10 14MHz)
ADC	Octal channel 12-bit, 65 MSPS

A chip TX810 [11] of Texas Instruments was selected as the T/R Switches. In Table 3 is shown its main electrical characteristics.

Table 3. Main Electrical Characteristics of TX810

Parameter		Value
Maximum Power Dissipation		644 mW
Input Amplitude	1 μs positive and negative pulse	+/-90 V
	continuous wave	+/-10 V
Insertion loss		from -0,9 to -7 dB
-3dB Bandwidth		from 65 MHz to 140 MHz
Input Referred Noise		0,91 – 1,12 nV/√Hz

The AFE5851 and the TX810 are dedicated to ultrasound applications using phased array transducer. They allow to implement complete receive path of the device.

The theoretical limit of the input echo-signal maximum frequency is limited by ADC and it is just over 16 MHz. The echo-signal at frequency of 9 MHz digitized by AFE5851 is shown in Figure 2.

ADC sampling frequency can be adjusted by changing frequency of the external oscillator connected to AFE5851. Therefore it has been chosen crystal controlled oscillator with frequency switch capability.

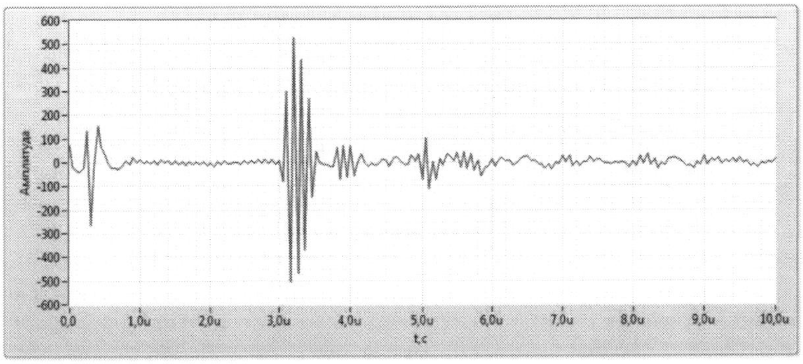

Figure 2. The echo-signal at frequency of 9 MHz digitized by AFE5851.

The AFE5851 output signal is passed to the FPGA via LVDS interface. All LVDS interface paths must have the same length. It should be taken into consideration during designing the printed circuit board.

LM96550 [9] и LM96570 [12] of National Semiconductors Corporation was selected as the Transmit Pulser and the Transmit Beamformer. The main electrical characteristics of chips is shown in Table 4.

Table 4. Main Electrical Characteristics of LM96550 and LM96570

Parameter		Value
Number of channels		8
Beamformer Delay Resolution		0.78 ns
Beamformer Delay Range		102.4 us
Sequences		64 pulses
1σ Output Jitter		25 ps
Output Frequency Range		1-15 MHz
Output Current	2% Duty Cycle	2 A
	100% Duty Cycle	0.6 A
Output Voltage		±(3.3-50) V

The advantages of the used chipset are:
- They are multi-channel chips, the possibility of increasing the number of channels by increasing the chip number;
- They can be relatively easy mounted on a printed circuit board since chip terminals are located on the edge of the chip, and there are no terminals on the chips bottom;
- High reliability of the chips. The chips have successfully passed all the stages of hardware debugging.

AFE5851 and TX810 have a fairly simple power supply scheme. These chips do not require establishing the supply voltage switching sequence. This advantage is not presented to LM96550 and LM96570. They require a large number of supply voltages (7 different values), that should be applied in strong specific sequence. The special control circuit should be used to control this sequence. Therefore, if it is necessary to adjust probe pulses amplitude, one has to change two stabilized voltage simultaneously.

Four layers PCB was developed and produced for receiving and transmitting units. One of the inner layer is set for ground connections. Another inner layer is used for supply voltage connection.

While implementing FPGA unit it is convenient to use commercially available prototyping board. It has been used the Digilent Genesys™ Virtex-5 FPGA Development Board [13]. Such a board has all required interfaces LVDS, USB, 100Base-T, and power supply stabilizers. Use of ready-made FPGA processing boards allows to save development time and resources.

4. Conclusions

Thus, it has been developed a 32-channel phased array ultrasound probing device. The work is performed according to the Russian Government Program of Competitive Growth of Kazan Federal University. The combination of the self-developed receiving and a transmitting units and commercially available FPGA unit and Personal Computer can solve our scientific goals, while providing a relatively low device cost.

References

[1] Denisov E S, Sagdiev R K, Evdokimov Yu K, Temyanov B K, Fazlyyyakhmatov M G 2013 *Proc. Int. Sci. and Tech. Conf. Nigmatullin's Readings -2013* (Kazan: Kaz. St. Tech. Un. Press) 134-137
[2] Temyanov B K, Evdokimov Yu K 2012 Vestnik of Tupolev KSTU **4 (2)** 121-125
[3] Temyanov B K, Evdokimov Yu K 2013 Vestnik of Tupolev KSTU **3** 55-59
[4] Gavrilova V A, Fazlyyyakhmatov M G and Kashapov N F 2013 J. Phys.: Conf. Ser. **479** 012010
[5] Gavrilova V A, Kashapov N F, Kashapov R N 2011 Biomedical Engineering **45** 198-200
[6] Galyautdinov R T, Kasparov N F, Luchkin G S 2003 Inzhenerno-Fizicheskii Zhurnal **75 (5)** 170-173
[7] Fadeev S A and Kashapov N F 2013 J. Phys.: Conf. Ser. **479** 012009
[8] Moles M 2007 *Advances in Phased Array Ultrasonic Technology Application: R/D Tech Guideline* (Waltham: Olympus NTD) p 492
[9] *LM96550 Ultrasound Transmit Pulser Datasheet* 2014: http://www.ti.com/lit/gpn/lm96550
[10] *AFE5851 16 Channel Variable Gain Amplifier (VGA) with Octal High Speed ADC Datasheet* 2014: http://www.ti.com/lit/gpn/afe5851
[11] *TX810 8-Channel, Programmable T/R Switch for Ultrasound Datasheet* 2014: http://www.ti.com/lit/gpn/tx810
[12] *LM96570 Ultrasound Configurable Transmit Beamformer Datasheet* 2014: http://www.ti.com/lit/gpn/lm96570
[13] *Genesys™ Virtex-5 FPGA Development Board Datasheet* 2014: http://digilentinc.com/Data/Products/GENESYS/Genesys_RM_VC.pdf

Acoustic sensing method based on the inverse problem of recovering the density profile and the bulk modulus of the inhomogeneous medium

Yu K Evdokimov[1], **B K Temyanov**[1] **and M G Fazlyyyakhmatov**[2]
[1] Kazan National Research Technical University named after A.N. Tupolev, 31 K. Marksa str., Kazan 420111, Russian Federation
[2] Institute of Physics, Kazan Federal University, 18 Kremlyovskaya str., Kazan 420008, Russian Federation

Email: bulat_tk@mail.ru

Abstract. The model of inhomogeneous acoustic medium without loss in dimensional approximation based on the Riccati equation for the impedance of the medium is considered. The inverse operator problem of recovering the coefficients of the Riccati equation is formulated. The numerical algorithm for solving the problem based on the integral equation coupling input acoustic impedance or admittance to the distributions of density or bulk modulus of the medium is proposed. The method of acoustic scanning is proposed. The numerical simulation of the proposed algorithm is carried out.

1. Introduction

In this paper we consider the problem of determining continuously varying mechanical properties of one-dimensional acoustic medium, in particular, the density profile or the bulk modulus that is relevant to NDT methods. It should be noted that most of the existing methods are based on measurement of irregularities in the form of abrupt disorders of the mechanical properties of the tested materials [1-3]. The proposed method will be tested and used on ultrasonic scanning system is being developed by our research group [4-6].

2. Mathematical model the acoustic medium in one-dimensional approximation

Propagation of acoustic oscillations in the medium is described by equations of hyperbolic type [7,8], which in a frequency domain are the system of equations:

$$\frac{\partial p(x, j\omega)}{\partial x} + \rho(x) j\omega v(x, j\omega) = 0, \tag{1}$$

$$\frac{\partial v(x, j\omega)}{\partial x} + \frac{j\omega}{k(x)} p(x, j\omega) = 0, \tag{2}$$

where $\rho(x)$, $k(x)$ are the density and bulk modulus of the medium [9], $p(x, j\omega)$ and $v(x, j\omega)$ are the pressure and particle velocity in the medium [9]. We make the transition to the impedance of the acoustic medium:

$$Z(x, j\omega) = \frac{p(x, j\omega)}{v(x, j\omega)}, \tag{3}$$

and the system (1, 2) we transform to the Riccati equation [7]:

$$\frac{dZ(x, j\omega)}{dx} - \frac{j\omega}{k(x)} Z^2(x, j\omega) + j\omega\rho(x) = 0, \tag{4}$$

where $x \in [0, l]$.

Analogous equation can be derived for the admittance of the medium:

$$Y(x, j\omega) = \frac{v(x, j\omega)}{p(x, j\omega)}, \tag{5}$$

$$\frac{dY(x, j\omega)}{dx} - j\omega\rho(x)Z^2(x, j\omega) + \frac{j\omega}{k(x)} = 0. \tag{6}$$

3. Formulation of the inverse problem

We form the equation Riccati which generalizes equations (4) and (6):

$$\frac{dH(x, j\omega)}{dx} - j\omega\gamma(x)H^2(x, j\omega) + j\omega\beta(x) = 0. \tag{7}$$

where $H(x, j\omega)$ is the immittance equal to impedance or admittance of the medium, $\gamma(x)$ and $\beta(x)$ for the equation (4) respectively equal $\frac{1}{k(x)}$ and $\rho(x)$, and for equation (6) equal to the same values, but in reverse order.

In this work the task of determining the coefficient $\beta(x)$ equation (7) for a known and given constant coefficient $\gamma(x)$ is assigned:

$$\gamma(x) = \gamma_0. \tag{8}$$

We formulate the inverse operator problem of determining $\beta(x)$ based on the equation (7).

As boundary conditions we consider the case when at the end of the line immittance equal to the definite value H_l:

$$H(l, j\omega) = H_l. \tag{9}$$

We also assume that the input immittance of the medium $H_S(j\omega)$ is measured with the standard error δ_S:

$$H(0, j\omega) = H_S(j\omega). \tag{10}$$

We believe that the small perturbation the density of the medium $\delta\beta(x)$ leads to the corresponding small increment immittance $\delta H_s(0,\mathrm{j}\omega)$. This relationship is represented as the Fredholm integral equation of the first kind [10,11]:

$$\int_0^l \exp\left(-\int_0^x j\omega\gamma_0 2H(x',j\omega)dx'\right)j\omega\delta\beta(x)dx = \delta H_S(0,j\omega). \qquad (11)$$

Solution of the integral equation Fredholm of the first kind is ill-posed problem. One of the standard methods for solving this equation is the Tikhonov regularization [7,10,11].

Using the boundary value problem (7), (9), by the relations (8), (10), (11) to recover the distribution of the parameter $\beta(x)$ is necessary.

4. The algorithm for solving the inverse problem and its numerical simulation

Based on the equation (4) the algorithm for determining $\beta(x)$ on the measured input immittance $H_s(\mathrm{j}\omega)$ is constructed:

1. Setting the initial approximation $\beta_0(x)$ and starting the iteration process ($i=0$ – number of iteration);
2. Calculation of the frequency response of immittance $H_i(x,j\omega)$ by solving the equation (7) for the model with the function $\beta_i(x)$;
3. Finding the right side of equation (11) by the formula: $\delta H_S(0,j\omega) = H_S(j\omega) - H_i(0,j\omega)$;
4. Calculation $\delta\beta_i(x)$ by solving the equation (11) using the method of Tikhonov regularization;
5. If $\int_0^l |\delta\beta_i(x)|^2 dx \le \varepsilon^2$, where ε is the set value, then $\beta_i(x)$ is the sought value, if not, then calculating the following approximation $\beta_{i+1}(x) = \beta_i(x) + \delta\beta_i(x)$ with the transition to step 3.

The reconstruction medium density $\rho(x)$ algorithm from the measured input impedance $Z_s(j\omega)$ and the algorithm for determining the bulk modulus $k(x)$ on the input acoustic admittance $Y(0,j\omega)$ was considered for the numerical realization. Main relations (7), (9) - (11) were reduced to dimensionless form: $\overline{x} = x/l$, $\overline{\omega} = \omega l\sqrt{\frac{\rho}{K_0}}$, $\overline{H} = H/j\omega\gamma_0$, $\overline{\beta} = \beta/\beta_0$, $\overline{\gamma} = \gamma/\gamma_0 = 1$. Riccati equation and the corresponding boundary conditions thus take a normalized form:

$$\frac{d\overline{H}}{d\overline{x}} - \overline{H}^2 - \overline{\omega}^2\overline{\beta(x)} = 0; \qquad 0 < \overline{x} < 1, \qquad (12)$$

$$\overline{H(0,j\omega)} = \overline{H}_l. \qquad (13)$$

Further, to simplify writing the upper horizontal line will be dropped and all statement in this article will be leaded to normalized values.

For the regularized solution of Fredholm integral equation of the first kind we used the standard program TIKH 1 given in [12], where the method of Tikhonov regularization with the regularization parameter α_T selection method of the generalized discrepancy is used. In all calculations we used the first-order regularizer. We set the upper α_{T1} and lower α_{Tm} bounds of the regularization parameter varies as:

$$\alpha_T > 0, \quad \alpha_{Ti} = \theta\alpha_{T(i-1)}, \; i = 2,3,....m, \quad 0 < \theta < 1. \tag{14}$$

As the lower bound α_m was selected value $\alpha_m \cong \delta_R^2$, where δ_R measurement error, and upper bound is was selected value $\alpha_T = 0{,}1$. Numerical solution of the Riccati equation was based on recurrent reverse BR - algorithm [10,13]. When solving the Fredholm integral equation of the first kind (11) was set error right side δ_R:

$$\left\|\tilde{H}_S - H_S\right\|_{L_2} = \left\|\delta H_S\right\|_{L_2} \le \delta_R, \qquad \delta_R = \delta_S\sqrt{\omega_{\max} - \omega_{\min}}; \tag{15}$$

where δ_S is the standard error of measurement of the real and imaginary parts of immittance H_S. Dependence coefficient $\beta(x)$ from the longitudinal coordinate is modeled in the form:

$$\beta(x) = \beta_0\psi_F(x). \tag{16}$$

As an example, the inhomogeneity function $\psi_F(x)$ was defined by two quadratic polynomials:

$$\psi_{F1} = 1 + 0{,}1x + 0{,}1x^2. \tag{17}$$

$$\psi_{F2} = 1 - 0{,}1x - 0{,}1x^2. \tag{18}$$

Probing the medium was carried out from one end. Area of solving the integral equation (11) and the Riccati equation (7) was set on the mesh $N_x \times N_\omega$, homogeneous frequency and spatial coordinates x.

Numerical experiments was carried out on PC using the programming environment Lab VIEW 8.5. Calculations were performed with complex numbers with double precision. The dimensionless frequency were lying in two ranges $0.032 < \overline{\omega} < 1.18$, $0.032 < \overline{\omega} < 1.24$ for expressions (17) and (18) respectively. In these frequency ranges error recovery is minimal. Space-frequency mesh dimension was $N_x \times N_\omega = 30 \times 100$. Boundary value of the immittance at the end of the line is zero:

$$H_l = 0. \tag{19}$$

In all examples, the initial approximations $\psi_F(x)$ was set constant and equal to unity.

In the Figure 1a and Figure 1b are shown respectively desired (1) and reconstructed (2) profile inhomogeneity function $\psi_{F1}(x)$, defined in the form (17), when the measurement error input immittance $\sigma_S = 0.1\%$, 1%. The relative error recovery amounted 0.94%, 1.04%.

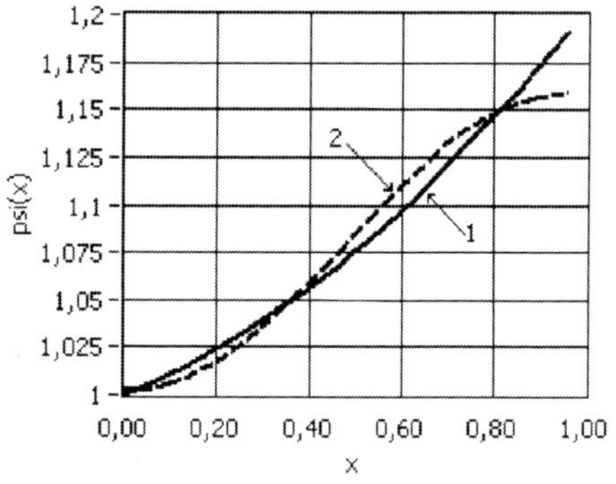

Figure 1a. Desired profile inhomogeneity function $\psi_{F1}(x)$

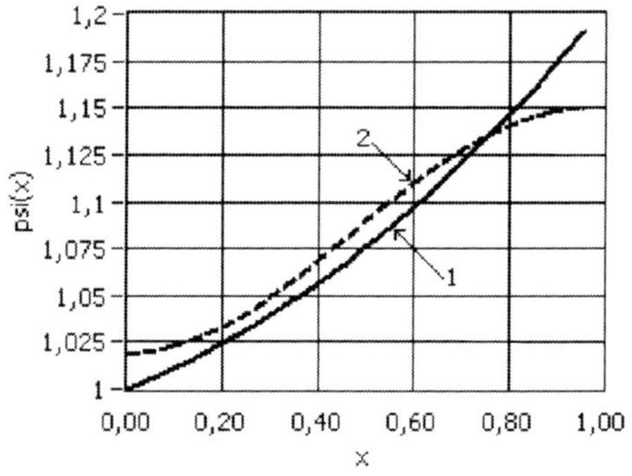

Figure 1b. Reconstructed profile inhomogeneity function $\psi_{F1}(x)$

In the Figure 2a and Figure 2b are shown respectively results of restoration the function $\psi_{F2}(x)$ defined in the form (18) at various levels of error σ_S (0,1%, 1%) measurements of input immittance $Z_S(j\omega)$. For this case, the error of the algorithm was amounted 1.21%, 2.09%, respectively. Calculations showed that the first 2-3 iterations provide a very rapid approach to the reversal of the desired curve, and then the process of continuous improvement solutions comes with a very low speed.

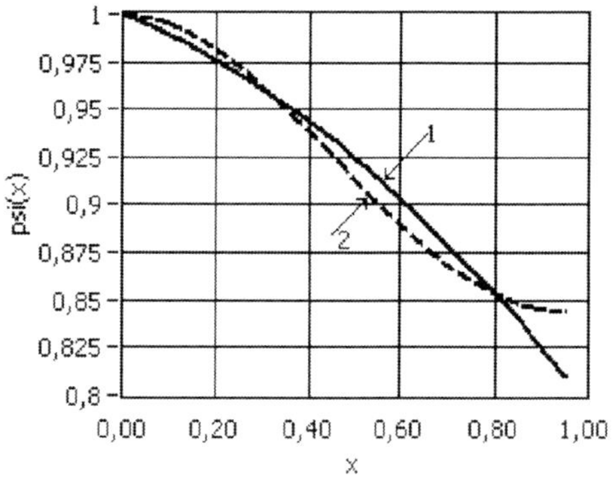

Figure 2a. Result of restoration function $\psi_{F2}(x)$ at level of error σ_S=0,1%

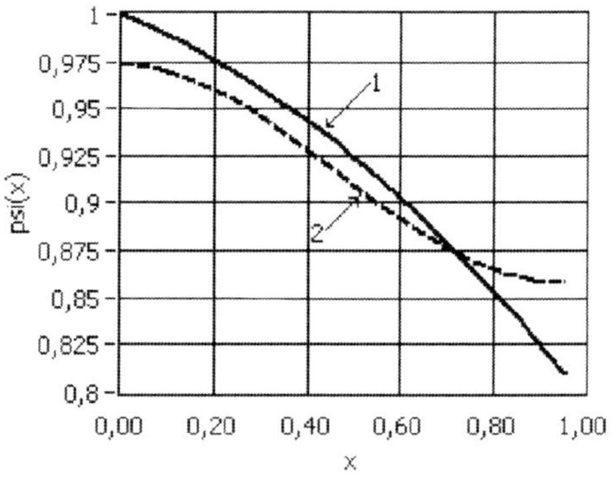

Figure 2b. Result of restoration function $\psi_{F2}(x)$ at level of error σ_S=1%

5. Conclusions

In this work the model of inhomogeneous acoustic medium without loss in dimensional approximation based on the Riccati equation for the impedance of the medium is considered. The inverse operator problem of recovering the coefficients of the Riccati equation is formulated. The numerical algorithm for solving the problem based on the integral equation coupling input acoustic impedance or admittance to the distributions of density or bulk modulus of the medium is proposed. The method of acoustic scanning is proposed.

References

[1] Barkhatov V A 2003 Russian Journal of Nondestructive Testing **39** (**1**) 23-47

[2] Agayan G M, Vinogradov N S, Goncharskij A V, Ovchinnikov S L and Romanov S Yu 2007 Vychislitel'nye Metody i Programmirovanie **8** 154-159
[3] Samokrutov A A and Shevaldykin V G 2011 Russian Journal of Nondestructive Testing **47 (1)** 16-29
[4] Denisov E S, Sagdiev R K, Evdokimov Yu K, Temyanov B K and Fazlyyyakhmatov M G 2013 *Proc. Int. Sci. and Tech. Conf. Nigmatullin's Readings -2013* (Kazan: Kaz. St. Tech. Un. Press) 134-137
[5] Gavrilova V A, Kashapov N F and Kashapov R N 2011 Biomedical Engineering **45** 198-200
[6] Gavrilova V A, Fazlyyyakhmatov M G and Kashapov N F 2013 J. Phys.: Conf. Ser. **479** 012010
[7] Kabanikhin S I 2009 *Obratnye i Nekorrektnye Zadachi* (Novosibirsk: Siberian scientific publishing) p 457
[8] Evdokimov Yu K, Ishuev T N and Khramov L D 2007 *Proc. Int. Sci. Conf. Mathematic methods in technique and technology (MMTT-20)* (Yaroslavl': Yar. St. Tech. Un. Press) 76-78.
[9] Solovyanova I P and Shabunin S N 2004 *Teoriya volnovyh processov: Akusticheskie volny* (Ekaterinburg: Ur. St. Tech. Un. Press) p 142
[10] Evdokimov Yu K, Martemianov S 2004 Int. J. Heat Mass Transf. **47 (2)** 329-340
[11] Tikhonov A N and Arsenin V Ya 1979 *Metody Resheniya Nekorrektnyh Zadach* (Moscow: Science) p 284
[12] Verlan' A F and Sizikov V S 1986 *Integral'nye Uravneniya: Metody, Algoritmy, Programmy* (Kiev: Naukova Dumka) p 544
[13] Jones U and Tron V 1985 *Nepreryvnye Drobi: Analiticheskaya Teoriya I Prilozheniya* (Moscow: World) p 414

Beam Control System for Ultrasound Scanning Device

E S Denisov[1], B.K. Temyanov[1], R K Sagdiev[1] and M G Fazlyyyakhmatov[2]
[1] Kazan National Research Technical University named after A.N. Tupolev, 31 K. Marksa str., Kazan 420111, Russian Federation
[2] Institute of Physics, Kazan Federal University, 18 Kremlyovskaya str., Kazan 420008, Russian Federation

Email: mfazlyjy@kpfu.ru

Abstract. The problems of FPGA based beam control system for ultrasound scanning system is considered. As a result of enhancement of algorithm of linear phased array focusing method for increasing efficiency of phased array elements is proposed. The application of the method for the real ultrasound scanning system is also considered.

1. Introduction

Ultrasound scanning systems (USS) is one of the most effective modern nondestructive testing tools which allow to carry out reliable detection and accurate localization of the defects. Such systems can be configured to obtain narrow linear or focused beam which can be moved in some region to scan the object under test.

Currently the most effective USS are based on ultrasound phased arrays (PA) which usually consist of piezoelectric crystal elements placed in the polymer insulating matrix [1,2]. The main feature of phased array ultrasonic technology is the computer-controlled excitation of individual elements in multielement probe. The excitation of the elements allows generating an ultrasound focused beam the parameters of which such as angle, focal distance, and focal spot size can be modified in wide ranges. The beam parameters are controlled by the delays (phases) and amplitudes of each element in PA.

The delays distribution over the PA elements corresponding to the required beam parameters is specified by a certain focal law. The focal law for probes which is in direct contact with the test piece and intended to generate longitudinal waves has a parabolic shape [3]. Then the large scanning angle is used some elements of PA becomes not active that leads to decreasing efficiency of USS. To eliminate this problem the original algorithm has been proposed. This algorithm provides correction of focal law to activate all array elements and increase efficiency of USS. To implement this algorithm the special FPGA controlled USS has been developed.

2. Improved algorithm of linear phased array focusing

Convenient algorithm for ultrasound beam focusing uses the focal law of parabolic shape which is rotated in polar system of coordinates [4] (Figure 1). The general expression for this shape has the following form:

$$Ax^2 + By^2 + Cxy + Dx + Ey + F = 0, \quad (1)$$

$$y = v\tau, \quad (2)$$

$$A = \sin^2\theta / 2f\,, \tag{3}$$

$$B = -\cos^2\theta / 2f\,, \tag{4}$$

$$C = -\sin(2\theta) / 2f\,, \tag{5}$$

$$D = \sin(\theta)\,, \tag{6}$$

$$E = -\cos(\theta)\,, \tag{7}$$

$$F = p^2(N-1)^2 / 4f\,, \tag{8}$$

where τ is time delay, v is the sound velocity in the medium, f is the focal length, N is the number of elements, p is the size of the single element, θ is an angle in the polar coordinate system.

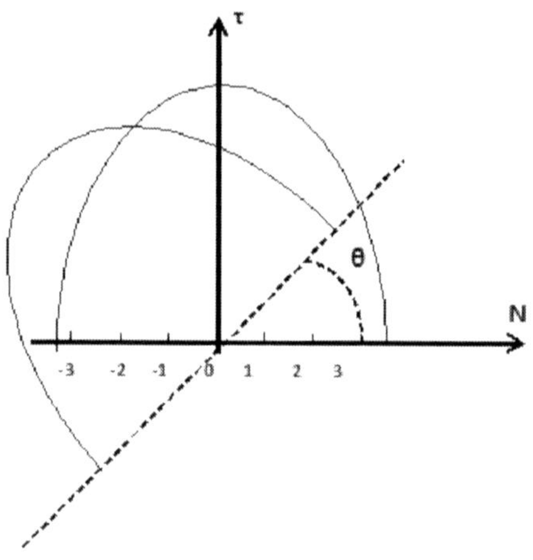

Figure 1. The focal law for one-dimensional phased array: τ is delay, N is a number of element, θ is rotation angle.

For large values of the scanning angle the boundary elements of array are not involved since they are leaving the parabola field. Thus in Figure 1 the right element is not involved due to turning. For the most efficient use of PA the displacement of parabola in direction of focal axis is proposed. For this purpose the focal law is changed. The focal distance is also should be adjusted so that the focal point (Figure 2) remains unchanged. It allows to activate all elements of PA. To keep the focal point in the predetermined position the condition of constant radius vector should be fulfill (Figure 2):

$$|r| = const\,. \tag{9}$$

The adjusted focal length could be determined from the equation:

$$f_1^2 - |r|\,f_1 - \frac{p^2(N-1)^2\cos(\theta)^2}{8} - \sin(\theta)\frac{p(N-1)}{2}f_1 = 0. \tag{10}$$

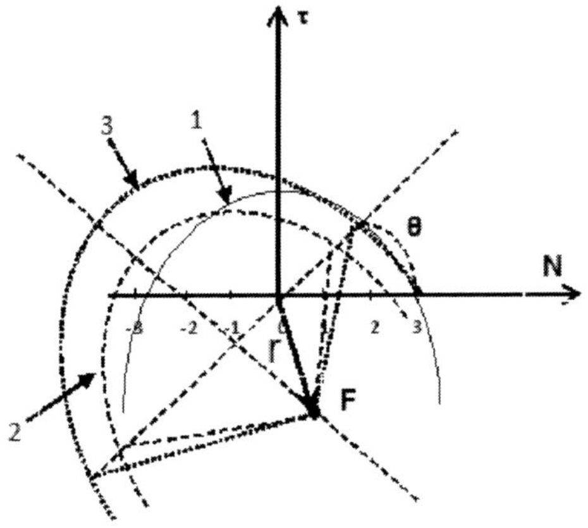

Figure 2. Correction of one-dimensional phased array focal law: τ - delay, N – elements number, θ - rotation angle, F- focus point, r – radius vector, 1 – non-deployed parabola, 2 – parabola rotated by the angle θ, 3 - adjusted deployed parabola.

In equation (1) the coefficient F (8) is changed after this correction:

$$F = p^2(N-1)^2 / 8f + \sin(\theta)p(N-1)/2. \tag{11}$$

This corrected focal law allows to increase efficiency of the PA.

3. FPGA based ultrasound scanning system

One of the most important, expensive and technically complex element of ultrasound scanning system is digital control unit. The main purposes of this system are beam forming and primary experimental data processing. Though the digital control system was developed to control ultrasound scanning system described in [5] it can be adapted to a wide range of the similar type systems.

The ultrasound scanning system [5] comprises the transmitting and receiving parts. The transmitting part consists of a transmit beamformer (LM96570 [6] of National Semiconductors Corporation) generating exciting pulses with individually controlled delays and transmit pulser (LM96550 [7] of National Semiconductors Corporation) forming the amplitudes of the pulses and the ultrasound transducer converting the electrical pulses into the ultrasound waves with the required parameters. The receiving part consists of the following main parts: the transmit/receive switches (TX810 [8] of Texas Instruments) preventing passing the large exciting pulses to the input of high sensitivity elements, the time variable gain amplifier, low pass filter and analog-to-digital converter (ADC) realizing using a single IC AFE5851 [9] of Texas Instruments. This ultrasound scanning system is controlled by the special dedicated digital system based on FPGA IC.

The digital control system should meet the following requirements:

1. Beam forming control for 32 channel ultrasound transmitter, specifically it should control delays and amplitudes of exciting pulses for each transmitting element;
2. Ultrasound waves transmitting triggering;
3. High speed ADC configuring and triggering;
4. Receiving sampled signal from ADC;
5. Preprocessing of the obtained experimental data.

Analysis of the requirements shows that the strongest limitations is caused by receiving data from 32 channels of high speed ADC. Indeed if the ADC with sampling frequency of 32MHz (for each channel) and resolution of 12 bit then taking into account that number of channel is equal to 32, the data rate of 12GS/s is required. Implementation of so high data rate of receiving signals with the requirements of real time referencing is extremely difficult operation. Currently the most efficient way to implement this operation is usage of Field Programming Gate Array (FPGA) based devices since they have well-developed efficient parallel architecture. It has been used Virtex-5 FPGA [10] of Xilinx Corporation. To implement high rate data receiving from the multichannel ADC the LVDS interface has been used.

Besides FPGA the digital control system also comprises: clock oscillator, power supply, ROM for FPGA configuration file storing and RAM for pulse forming and data processing assistance, interfaces for connection with the abovementioned components of the transmitting and receiving parts, and with computers and others FPGAs.

The block diagram of the digital control system is presented in Figure 3.

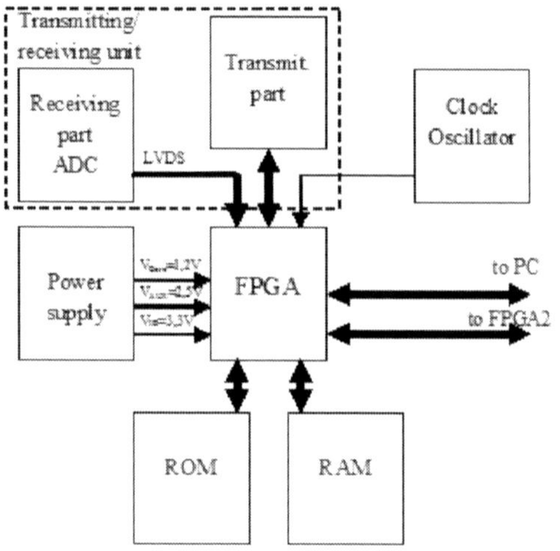

Figure 3. Block diagram of the digital control system.

During operation FPGA generates a control signals for transmitting and receiving parts. FPGA defines focal law and the ultrasound beam parameters. Specifically FPGA implements the proposed above improved algorithm of linear phased array focusing which allows to increase efficiency of the ultrasound phased array. After implementing of the proposed algorithm FPGA configures the beamformer and the pulser to obtain the required ultrasound waves. Then FPGA generates signal to start measurement and receives the signals from ADC.

The developed system is adapted to operate under control of a computer. The special software provides the following possibilities: 1) Control of the ultrasound scanning system by operator; 2) Graphical representation of the actual hardware configuration and results of measurements; 3) Transmitting control information to FPGA; 4) Receiving measurement results (experimental data); 5) Ultrasound scanning device technical state monitoring.

4. Conclusions

The FPGA based beam control system for ultrasound scanning system is developed. The work is performed according to the Russian Government Program of Competitive Growth of Kazan Federal University. It was shown that the correct selection of the pulses delays controlling beam parameters can increase efficiency of the ultrasound transceiver. The algorithm of linear phased array focusing and method of increasing efficiency of phased array elements are proposed and implementation of the proposed method for real ultrasound scanning system is considered. It has been shown that this method can be used for a wide range of ultrasound nondestructive testing equipment based on an ultrasound phased arrays.

References

[1] Gavrilova V A, Fazlyyyakhmatov M G and Kashapov N F 2013 J. Phys.: Conf. Ser. **479** 012010
[2] Gavrilova V A, Kashapov N F, Kashapov R N 2011 Biomedical Engineering **45** 198-200
[3] Moles M 2007 *Introduction to Phased Array Ultrasonic Technology Applications: R/D Tech Guideline* (Waltham: Olympus NTD) p 368
[4] Beklemishev D V 1998 *The Course of Analytical Geometry and Linear Algebra* (Moscow: Vyshaya shkola) p 350
[5] Denisov E S, Sagdiev R K, Evdokimov Yu K, Temyanov B K and Fazlyyyakhmatov M G 2013 *Proc. Int. Sci. and Tech. Conf. Nigmatullin's Readings -2013* (Kazan: Kaz. St. Tech. Un. Press) 134-137
[6] *LM96570 Ultrasound Configurable Transmit Beamformer Datasheet* 2014: http://www.ti.com/lit/gpn/lm96570
[7] *LM96550 Ultrasound Transmit Pulser Datasheet* 2014: http://www.ti.com/lit/gpn/lm96550
[8] *TX810 8-Channel, Programmable T/R Switch for Ultrasound Datasheet* 2014: http://www.ti.com/lit/gpn/tx810
[9] *AFE5851 16 Channel Variable Gain Amplifier (VGA) with Octal High Speed ADC Datasheet* 2014: http://www.ti.com/lit/gpn/afe5851
[10] *Genesys™ Virtex-5 FPGA Development Board Datasheet* 2014: http://digilentinc.com/Data/Products/GENESYS/Genesys_RM_VC.pdf

The analysis of the bottom electrode thermal streams during work process in a melting furnace

I K Israfilov, A T Galiakbarov, D I Israfilov
Kazan Federal University,
423800, 68/19 Mira avenue, Naberezhnye Chelny, Russia

E-mail: chelny@kpfu.ru

Abstract. Conducted extensive research of thermal work of the existing bottom electrode (BE) of copper in operation. Calculation of optimum parameters (water cooling, temperature of the walls and parameters) BE of copper under specific conditions. Compiled formulas and graphs to determine the optimal geometry of BE under different operating conditions.

1. Introduction

During the task of the set task it is accepted to observe BE cooling from the point of heat-transfer theory in which transfer of heat from one medium to another through divided body of any form includes a convective heat exchange from hotter heat-transfer agent to a body, heat conductivity in a body, a convective heat exchange from a body to colder medium [1].

2. The basic part

Under the set conditions it is necessary to find a thermal stream from a hot liquid to cold and temperatures on a surface of a wall [2].
Density of a thermal stream through a body will be equal to [1]:

$$q = \frac{t_{ж1} - t_{ж2}}{\frac{1}{\alpha_1} + \frac{\delta}{\lambda} + \frac{1}{\alpha_2}}$$

Thermal stream Q, Watt, through surface F of a firm wall is equal to:

$$Q = q \cdot F$$

It is necessary to apply similarity theory to define factors of a convective heat exchange α_1 и α_2 [3]. The data obtained during experiments is generalized by means of a similarity parameter system. For this purpose results of experiments on definition of convective heat exchange factors α are represented in a form of criterion equation [4].

$$Nu = CRe^m Pr^n Gr^p,$$

where $Nu = \alpha * l/\lambda$ is the dimensionless group named a Nusselt number; here α - an average on a flowing surface factor value of a convective heat exchange; l - a characteristic size; λ - factor

The convective heat exchange at boiling is connected with change of a modular condition. In a thin layer, at a heating up wall, bubble boiling is originated.
Film boiling on a heating surface forms a continuous film of steam [3,5].

Steam film creates an additional thermal resistance.

For exclusion of film boiling during the work of bottom electrode, it is offered to apply the helical chilling channel with a point of an entry of a coolant in the bottom of a spiral and an exit - to a top of a spiral that allows to move formed gas bubbles upwards on the chilling channel without a delay of bubbles to an exit from the bottom electrode.

Results of generalization of experimental data.

Indications at work of the small furnace on 250 kg for cast iron melting have been spent. A current 1900 A, voltage 83 Watt, melting temperature 1550^0 C, ambient temperature 10^0 C, cooling water temperature on an entry of the bottom electrode 25^0 C, on an exit 33^0 C, water discharge 2 m^3/h.

Being based on experimental data and using formulas on heat-transfer theory, we count temperatures on a bottom copper electrode surface from liquid melt and from cooling water. Thus, we gain a temperature drop on an electrode depending on a thermal resistance. Let's observe sequentially all resistance.

Thermal resistance at a convective heat exchange "liquid melt – BE".

The convective heat exchange of metal liquid melt to BE occurs due to a free convection.

Communication Nu at a free convection with a Grashof and Prandtl number looks like [4]:

$$Nu=0,105*Gr^{1/3}*Pr^n,$$

where $n=0.3+0.02/Pr^{1/3}$

If the number of Grashof is great ($>10^9$), then:

$$Nu=0,017*Gr^{1/3}*Pr^n,$$

where $n=0.3+0.02/Pr^{1/3}$

Heat conductivity thermal resistance BE.

It is defined by formula [1]:

$$R=\delta*F/\lambda,$$

where F is a cross-sectional area, δ - a thickness of a wall, λ - coefficient of thermal conductivity.

Thermal resistance at a convective heat exchange "BE – water".

The convective heat exchange from BE to water occurs due to a forced convection [6]. The Convective heat transfer occurs due to contact of a moving liquid and the firm surface, having various temperature. At a forced convection this process is going not by liquid heating as it is observed at a free convection, but by affecting of some superposed force. Influence of these forces on a liquid convective heat exchange is characterized by dimensionless parameter - a Reynolds number. This parameter directly defines a liquid convective heat exchange.

The convective heat exchange at a liquid forced convection in pipes, a cross-flow of pipes and along flat surfaces is investigated full enough. Results of these experimental researches are usually interpolated by Nusselt equation number [4]:

$$Nu=C*Re^m*Pr^n,$$

For water [4]:

$$Nu=0,035*Re^{0,8}*Pr^{0,4}$$

Using experimental data, BE temperature on walls is calculated from liquid melt T_{w1} and water T_{w2} taking into account embedded power, weight and a liquid melt chemical compound. Calculation was conducted in MathCAD environment. Also difference of temperatures ΔT of a coolant (water) on BE entry and exit is resulted. Calculations were conducted for the furnace on 250 which is used for cast iron and ferromolybdenum melting and 500 kg used for cast iron melting. The data is presented in table 1.

For the furnace (1000kg) at ferromolybdenum melting (Fe-Mo), working on old BE design a settled temperature T_{w1} makes 1500^0C.

Table 1. BE wall temperature value T_{w1} near liquid melt, BE T_{w2} near water and temperature difference ΔT of a coolant.

Furnace	250 kg.		500kg	1000kg
Material	Cast iron	Fe-Mo	Cast iron	Fe-Mo
R1, Watt/K	0,0177	0,0342	0,008	0,0079
R2, Watt/K	0,0350	0,0350	0,0190	0,0190
R3, Watt/K	0,0034	0,0034	0,0034	0,0042
Tw1, ^{0}C	1016	1057	1086	1500
Tw2, ^{0}C	119	123	193	300
ΔT ^{0}C	9	10	15	25

3.Results.

In connection with expected BE overheating that leads to erosion [7], it is recommended to increase quantity of heat output at the expense of creation of a ridge surface from a coolant.
Ribs in a cross-section can have a profile of the most various geometrical configurations (a rectangle, a circle, a triangle and other figures, including wrong geometrical form) [8, 9].
Fig. 1 represents the circuit design of a cylindrical rib. At calculation of rib sizes the ribbing factor E_{rib} is important which is defined according to the table [1] depending on convective heat exchange factor α, rib sizes (R1, R2 and δ) etc.

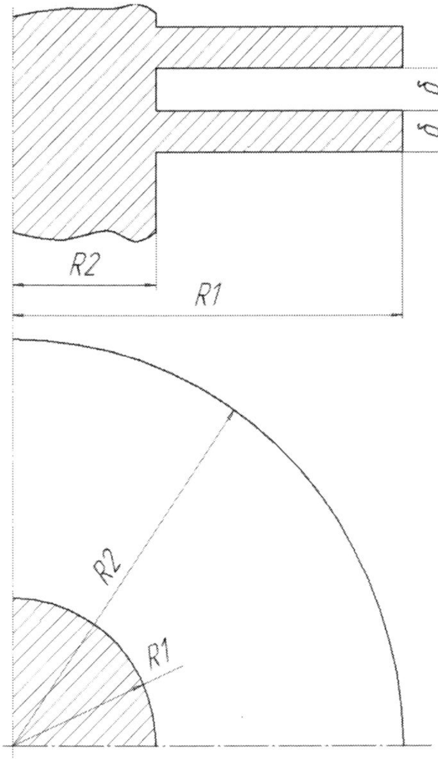

Fig.1. The circuit design of a flat round rib

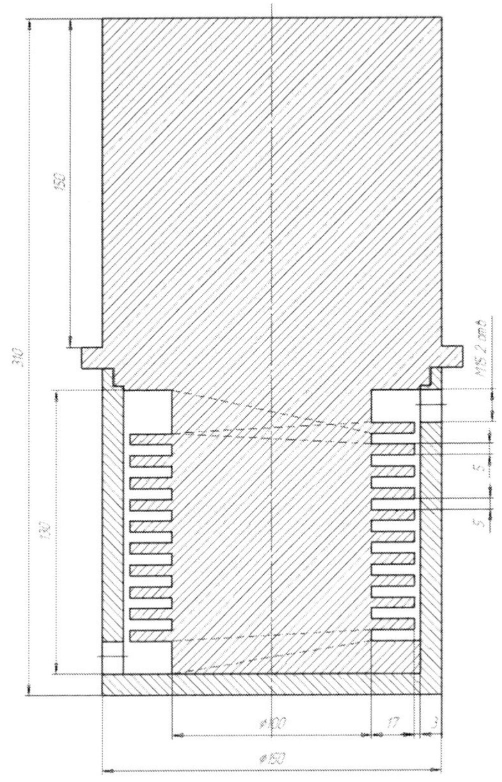

Fig.2. Recommended electrode

The thermal stream from ribbing surface is defined by formula [10]:

$$Q = \alpha_2 * F_{rib} * (T_1 - T_2) * E_{rib},$$

where F_{rib} - the ribbing square

T_1, T_2 - wall and liquid temperatures.

Let's make BE thermal calculation with the set sizes, applicable for the furnace with power of 1MWatt and ferromolybdenum melting temperature 2000^0C.

Calculation is conducted in MathCAD environment. During calculation the following design of an electrode is offered (fig. 2), allowing to increase duration of their work.

Conclusion: BE length changing without change of liquid melt contact square with an electrode is not recommended [11] as it (at constant contact diameter of liquid melt with BE) will lead to growth of R_2 and therefore, to temperature drop growth on the bottom electrode, thus from liquid melt difference of temperatures decreases, and the temperature of BE wall increases from liquid melt. That leads to BE flashing off and its subsequent falling out.

3. References

[1] Kuteladze S S 1990 *Heat transfer and flow resistance*: A Reference Guide. - Energoatomizdat , – 367 p

[2] Israfilov I H 2007 *Generalized criteria for estimating the parameters of concentrated energy fluxes in the development of automated process equipment* / Interuniversity scientific book "Design and study of technical systems " - Neb . Chelny Univ INEC. Number 11 p 88- 94

[3] Isachenko V P etc. 1981 *Heat*: Textbook for universities. – Energoatomizdat 416 p

IOP Conf. Series: Materials Science and Engineering **69** (2014) 012015 doi:10.1088/1757-899X/69/1/012015

[4] Uong H 1979 Basic formulas and data on heat transfer for engineers. Handbook . - Oxford: Clarendon Press , – 216 p
[5] Abdullin I Sh, Kashapov N F and Kudinov V V 2000 Fizika i Khimiya Obrabotki Materialov **3** 35-38
[6] Lopuhov G A 2004 *Bottom electrode arc furnace¬constant direct current design company DANIEL* / Electrometallurgy. Number 6 P 48
[7] Israfilov I H 2011 *The study traces the electric arc electrodes in a pulsed plasma generator* / Socio-economic and technical systems : electronic periodical. - Neb . Chelny № 2 (59)
[8] Rayle B 2010 Bottom electrode arc furnace SIMETALcis DC / / Electrometallurgy Number 12.pp 18-20
[9] Liu X 2008 *Melting mechanism of water-cooled billet-type bottom electrode of direct current arc furnace: a numerical approach* / Metallurgical and materials transactions B. - October. - № 39B – p 713 -724
[10] LI Roizen 1977 Thermal design of finned surfaces. M. Energy - With 256
[11] Zaytsev V A 2009 Bottom electrode DC arc furnaces / Contemporary Electrometallurgy N 2 p 3-8.

IOP Conf. Series: Materials Science and Engineering **69** (2014) 012016 doi:10.1088/1757-899X/69/1/012016

Applying 3D-printing technology in planning operations of cancer patients

L N Kashapov[1], A N Rudyk A.N.[2] and R N Kashapov[1,3]
[1] Engineering Institute, Kazan Federal University, Kazan, 420008, Russia
[2] Tatarstan Regional Clinical Cancer Center, Kazan, 420029, Russia
[3] Laboratory of Radiation Physics, Kazan Physical-Technical Institute of the Kazan Scientific Center of the Russian Academy of Sciences, Kazan, 420029, Russia

E-mail: kashramiln@gmail.com

Abstract. The purpose of this work was creation 3D model of the front part of the skull of the patient and evaluates the effectiveness of its use in the planning of the operation. To achieve this goal was chosen an operation to remove a tumor of the right eyelid, germinate in the zygomatic bone. 3D printing was performed at different peripheral devices using the method of layering creating physical objects by a digital 3D model as well as the recovery model of the skull with the entire right malar bone for fixation on her titanium frame to maintain the eyeball in a fixed state.

1. Introduction

Nowadays, 3D technology are increasingly being used in life, they are used in the design of new complicated profile products for various purposes. Special needs of additive manufacturing technology used in the methods of manufacture of prototypes, and where each piece is an individual, not serial. Medicine is an area in which introduced various innovations from completely different fields of science and technology [1]. And 3D prototyping is no exception.

The purpose of this work was creation 3D model of the front part of the skull of the patient and evaluates the effectiveness of its use in the planning of the operation [2]. 3D printing was performed at different peripheral devices using the method of layering creating physical objects by a digital 3D model as well as the recovery model of the skull with the entire right malar bone for fixation on her titanium frame to maintain the eyeball in a fixed state. And the whole procedure should be evaluated in terms of security and the possibility of its introduction into mass use.

2. Experimental

To create a 3D model of the skull were used computed tomography files of the patient. With the program 3D Slicer were processed CT data and create a volumetric model of the skull of the patient in STL format. To do this, first add the CT files in the program 3D Slicer. Then based on our files program builds 2D models in three dimensions. On the obtained models are designated the patient's bone is then is launched the subprogram Model Maker to create a 3D model of the skull, and the resulting model is already saved in STL format.

Obtained STL model is necessary to modify, remove unwanted items. For this we used the Geomagic Studio 2012. With this program have been removed not needed for operation planning parts of the skull: the upper part of the frontal bone; parietal, occipital, temporal, and mandibular bones.

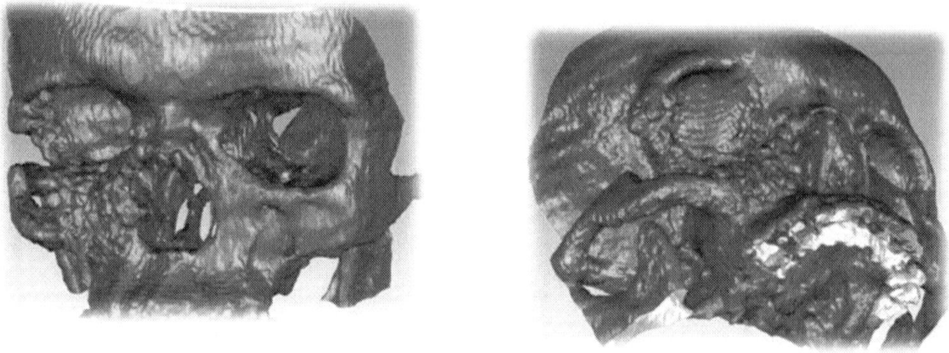

Fig.1. Obtained model after processing in the program Geomagic Studio 2012

The second point of our work was the creation a new 3D model of the patient's skull with a healthy right malar bone. We used a mirror operation of the program Geomagic Studio 2012 (Fig. 2). Then be used function of smoothing the surface, because model was based on CT, and the files are a definite step model and because of this it turns rough, stepped form (Fig. 3). And after all these operations the model data are saved in STL format.

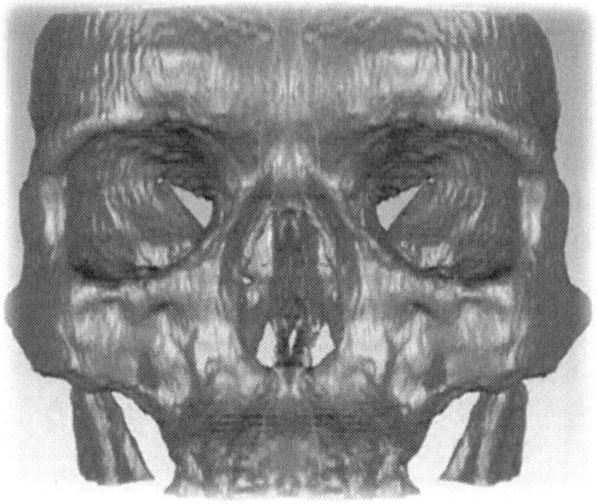

Fig.2 Mirroring of healthy (left) part of the skull on the right side

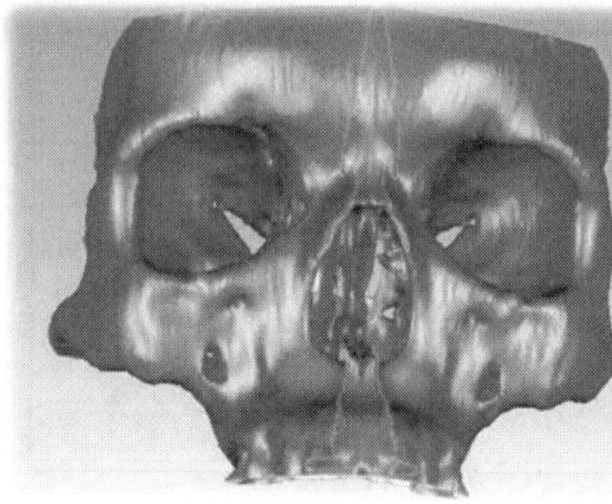

Fig.3 Smoothed model

Creating models produced on an industrial system 3D-printing Fortus 400mc and 3D CJP printer ProJet 160. These printers differ by technology of constructing the model, by the method of the overlay material and use a variety of materials of construction. Projet 160 uses as material powdery substance VisiJet PXL, and Fortus 400mc uses industrial-strength thermoplastics (ABS-M30, PC) with soluble and insoluble support.

For prototyping on the printer Projet 160 programs 3D Edit Pro and 3D Print used. The first program is used to correct defects in our model, which does not allow printing it on the printer Projet 160. And the 3D Print program is used to send the model to print. Printing time is 4-5 hours, and then 1-2 hours is cooling. Then a prototype is placed in a chamber for cleaning, which is undergoing blowing with compressed air. After the cleaning chamber for hardening the product, its tuck 10-15 seconds in a special binder solution ColorBond. Then a prototype is dried for about 2-3 minutes.

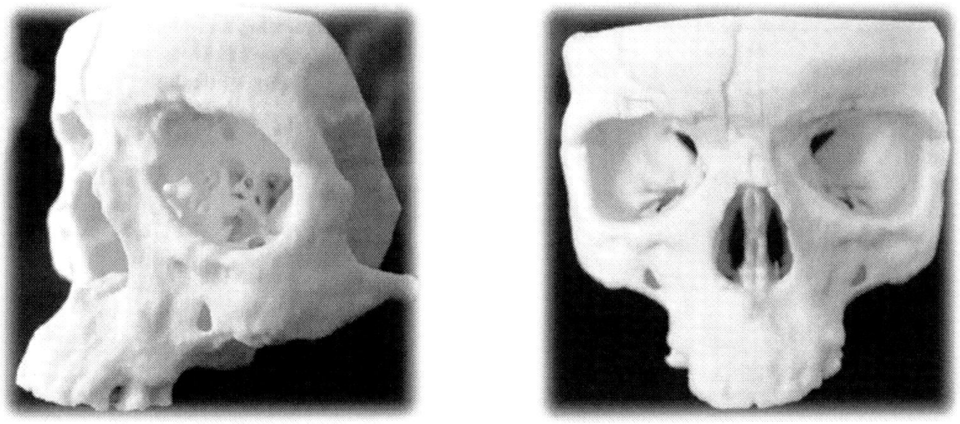

Fig.4. Assembled prototype of VisiJet PXL (Projet 160)

To create a prototype on industrial system of 3DprintingFortus 400mc used Insight and Control Center. The first program is used for programming the movement of the print head of the printer and creates G-code to the printer. The Control Center is positioning this prototype on the print area, and sends to print a model, and calculates the time of printing, the amount of material and support that is needed for printing. Print time is 14-15 hours and then for 1 hour is cooling. Then a prototype is placed in a chamber for treatment for 2-3 hours, where a special solution is removed by chemical support.

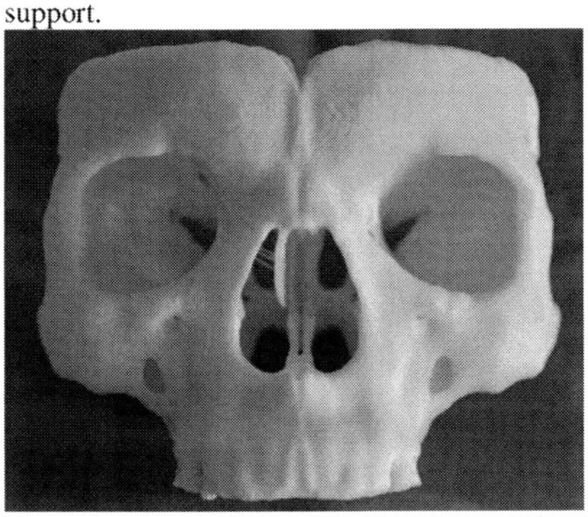

Fig. 5. Ready prototype of the PC (Fortus 400mc)

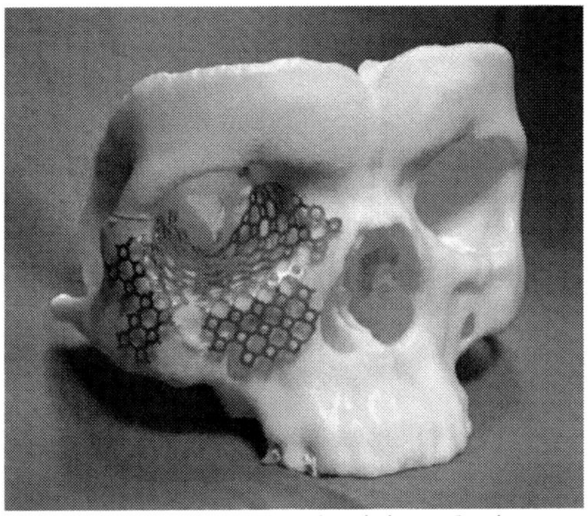

Fig.6. Overlay frame on the right malar bone and fixation.

Then plan to install individually simulated for the patient implant in the operational area of the defect, right maxilla and right orbit (Fig. 6). Now, this implant can be used for the operation after standard preparation. Application 3d prototyping technology in this case possible to reduce the operation time of 2-3 hours because simulation implant from the finished factory workpiece during operation requires more time and is associated with the risk of poor formation. By using 3D Slicer was created 3d model of the facial skeleton file-based CT scan. Then by using Geomagic Studio 2012, we edit the model under the required parameters, in this case, we cut part of the model is not used in operations.

3. Conclusions

It has been established that the difference from the existing methods is:

a) Prototyping and manufacturing functional anatomical models in less than 48 hours, at the same time providing a direct link clinics and laboratories.

b) There is the possibility of simulating the operation and preparation for any defect.

c) Becomes possible to modeling standard factory implants under the individual defect.

d) Reduction in the cost of the implant due to the pre-fitting of standard commercial products (grids, plates and so on.)

Acknowledgments

This work was financed by grants of Kazan (Volga region) Federal University of the state of tasks in the sphere of scientific activity.

References

[1] Gavrilova, V.A., Kashapov, N.F., Kashapov, R.N. Plasma application of protective polymer-powder coatings to ultrasonic sensors // Biomedical Engineering. - 2011 - 45 (5), pp. 198-200.

[2] Baj, D. ALI Youssef, R. Monteverdi, B. Bianchi, V.A. Combi, and A.B. Giannì. Reconstruction of partial maxillary defects with the double-barrel fibula free flap// Acta Otorhinolaryngol Ital. Dec 2010; 30(6): 299.

[3] B.J. Whyms, BS, H.K. Vorperian, L.R. Gentry, E.M. Schimek, E.T. Bersu, M.K. Chung The effect of CT scanner parameters and 3D volume rendering techniques on the accuracy of linear, angular, and volumetric measurements of the mandible. Oral Surg Oral Med Oral Pathol Oral Radiol. May 2013; 115(5): 682–691.

[4] James J. Xia, MD, Liza Shevchenko,Jaime Gateno, John F. Terry D. Robert E. Jeryl D. Chung H. Kathleen R. Outcome study of computer-aided surgical simulation in the treatment of patients with craniomaxillofacial deformities// J Oral Maxillofac Surg. Jul 2011; 69(7): 2014–2024.

IOP Conf. Series: Materials Science and Engineering **69** (2014) 012017 doi:10.1088/1757-899X/69/1/012017

Research of cobalt chromium alloy surface morphology after sandblasting

L N Kashapov[1], N F Kashapov[1] and R N Kashapov[1,2]
[1] Engineering Institute, Kazan Federal University, Kazan, 420008, Russia
[2] Laboratory of Radiation Physics, Kazan Physical-Technical Institute of the Kazan Scientific Center of the Russian Academy of Sciences, Kazan, 420029, Russia

E-mail: kashramiln@gmail.com

Abstract. The purpose of the work was to research of influence of sand blasting on surface morphology and properties of touching the metal-ceramic. Found that after the sand blasting on the metal surface are sand particles of size 100 - 200 nm are also observed in some places a large particle size of 2 - 3 mm, which is imprinted into the metal. It is shown that these particles can cause of crack nucleation at the metal-ceramic and lead to her cleavage.

1. Introduction

Sand blasting method is widely used in various industries. One of vivid examples is to use it in the methods of repairing and rebuilding worn machine parts [1]. Known products, which are based on metal, and on the surface is applied manually ceramic coating which is subsequently baked in thermal ovens. Previously metal treated with sand blasting detail when applying ceramic coatings by using plasma electric arc methods. Analysis of the statistics shows a large number of chips, not associated with structural features. This may be due to the properties of the contact points of the metal-ceramic. Therefore the purpose of work is investigating the effect of sandblasting on surface morphology and on the properties of touching the metal-ceramic.

2. Experimental

In industry mainly used products obtained by the method mechanical treatment or by casting processes. In our work for the approximation to the real conditions of the experiment, samples cobalt chromium alloy were cast on the fore-vacuum casting plant. For castings are manufactured originally a model cast tree on which wax patterns were fixed in the form of flat plates measuring 5 * 4 * 2 mm. At the second stage, the preparation and kneading the refractory molding composition Yeti Expansion with a liquid Yeti Liquid (22 ml per 50 grams). To avoid the formation of air bubbles using a vacuum mixer Fox 88 OMEC (Italy). Mixing time was 30 seconds, the speed of rotation blade 400 revolutions / minute, the pressure in the bowl - 0,8 Bar. Flooding cast tree produces on the vibro table VB 1.1 with 2 vibration frequencies 3000/6000 min^{-1}. After filling the molding material, waiting 30 minutes to complete solidification of the molding composition. The resultant shape of the molding composition was heated in a muffle furnace Programix 50 Ugin. This oven has a four-sided heating system that provides optimum heating and shape of the crucible with the metal. In our experiments we used the technique of shock heating of the casting-box. Namely, preheat the oven to a temperature of 900 ^{0}C. Cast-box was used to the size of the X6; the minimum warm-up time of the cast-box was 60 min. For efficient metal pouring crucible with the metal also heated in a muffle furnace. In the next step the heated crucible placed in a heating element of the casting installation. Produced heating metal with high frequency currents. Next, hot cast-box takes out from the muffle furnace and is attached vertically to the lid of the crucible and the from above is pressed by a pressing mechanism. After that the lid of the vacuum chamber of the casting installation is closed and evacuating air process begins with simultaneous high-frequency heating of the metal. In the process of heating the metal melts. Eventually melt of metal in the crucible is formed, which is coated with an oxide layer. At the moment of rupture of the oxide layer caster carries the casting process, which is as follows. Produced upend of vacuum casting chamber and pumping compressed air a speed of 3 bar / sec to a pressure of 3.6 bar as a result metal flows from the crucible into the cast-box and is compressed due to the pressure. Pressure allows pass of the metal in the canals to a thickness of 0.1 mm. after casting, vacuum casting chamber for 30 seconds is upside down, during which time there is metal cooling and crystallization. After that vacuum casting chamber comes in its original condition, then you can open the camera and remove the cast-box. After the cast-box had cooled necessary mechanically clean the metal frame from the molding composition. For research was selected cobalt chromium alloy firm Starbond CoS with the following composition and characteristics: Co - 59%, Cr - 25%, W - 9%, Mo - 3,5%, Si - 1%, another components (C, Fe, Mn, N) - max 1,5; ultimate tensile strength (Rp 0.2) 650 MPa, tensile strength 910 MPa, ultimate elongation 8, modulus of elasticity 200 GPa, Vickers hardness 280 HV 10, density 8,8 г/см^3, solidus-liquidus interval 1305 - 1400°C, casting temperature 1500 - 1550°C, thermal expansion coefficient 14,0 μm/м °C.

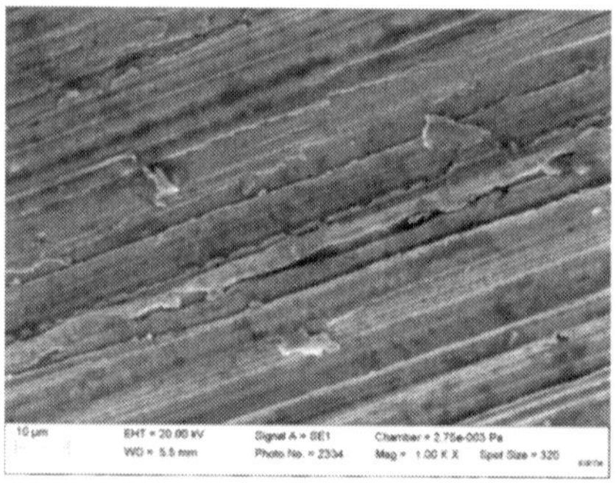

a - increase of 1,000 times

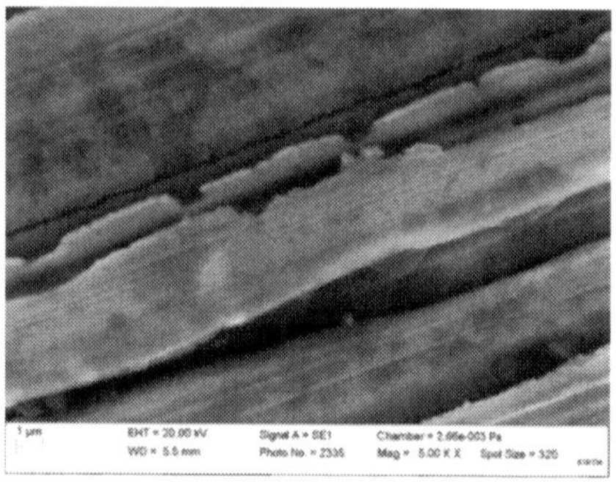

b- increase of 5,000 times

Fig. 1. SEM images of the surface after treatment carborundum disk.

In the final step was conducted mechanical treatment of the sample by carborundum disk. After mechanical treatment was performed treatment by a stream of aluminum oxide particles 99.5% Al2O3 and SiO2, not more than 0.06% grain size 110 - 150 microns (Cobra, Renfert) to install the POA-5U (Aveyron, Ekaterinburg). The surface morphology was investigated by scanning electron microscope EVO 50 Carl Zeiss AG. The elemental composition was determined by electron probe microanalysis using X-ray microanalysis system INCA Energy and INCA Wave. Profilogram surface roughness measured using a profilometer TR-200 [2]. Studies of the morphology and surface structure were performed before and after of sandblasting and surface treatment of the steam generator Omec GP92.3.

Figure 1 presents surface morphology of the samples after casting and polishing the final minutes. As can be seen the surface is flat and has a value of roughness R_a = 0,5 micron.

Figure 2 shows an image of the surface after of sandblasting, a - increase 1000 times b - 5000 times, c - 20 000 times. Figure 2 g shows profilogram of the surface. On it were defined the main parameters of roughness: R_a=1,75, R_y=11,13, R_t=15,23, R_m=5,41, R_z=8,23. The resulting surface has all the necessary parameters of the microrelief for the effective retention applied ceramic coatings [3]. However, it was found that after the of sandblasting on the metal surface stay sand particles of size 100 - 200 nm are also observed in some places a large particle size of 2 - 3 mm, which is imprinted into the metal. It is shown that these particles can cause of crack nucleation at the metal-ceramic and lead to her cleavage.

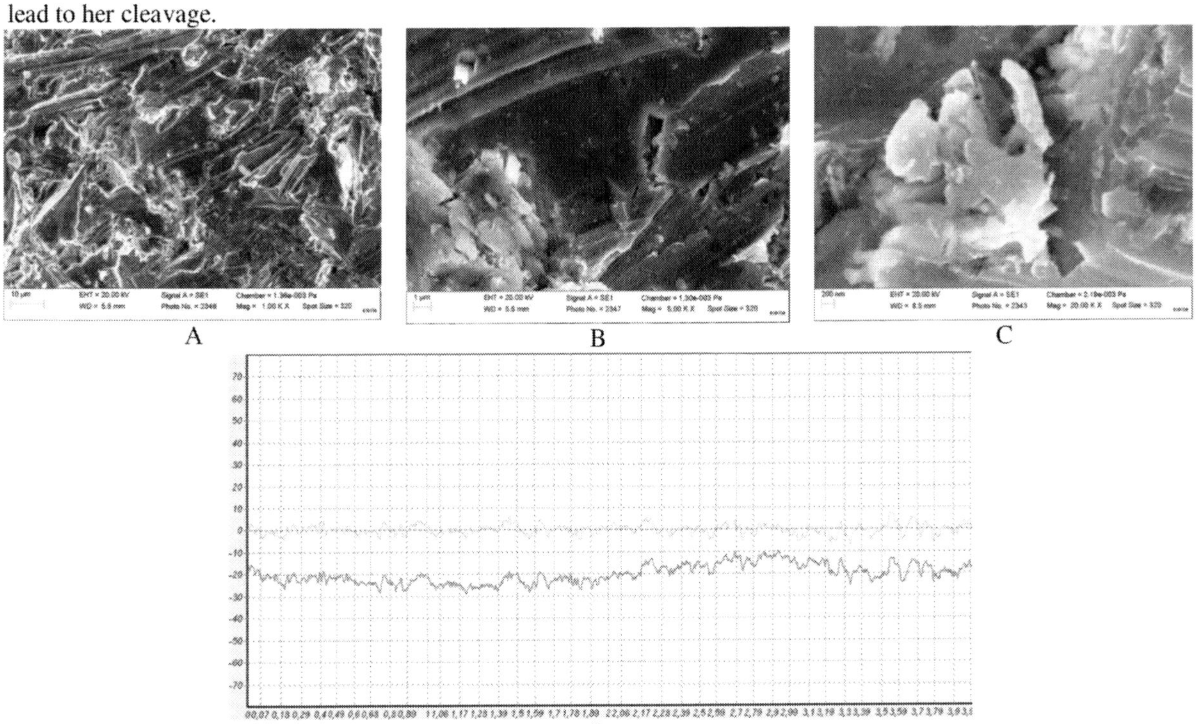

A B C

D

Fig. 2 Surface morphology after sandblasting: a - × 1000 b - × 5000, c - × 20000 g - profilogram of the surface.

Particles with nanometer range stay not possible to carry out the removal, without removal of the base metal. This will lead to a change in surface morphology. Studies were conducted to determine the composition of particles remaining on the metal surface. Figures 3a and 3b presents the place where analyzed. In Tables 1 and 2 set the resulting elemental composition. It is found that the particles are composed mainly of aluminum and oxygen, which corresponds to the composition of the used sand-Al_2O_3.

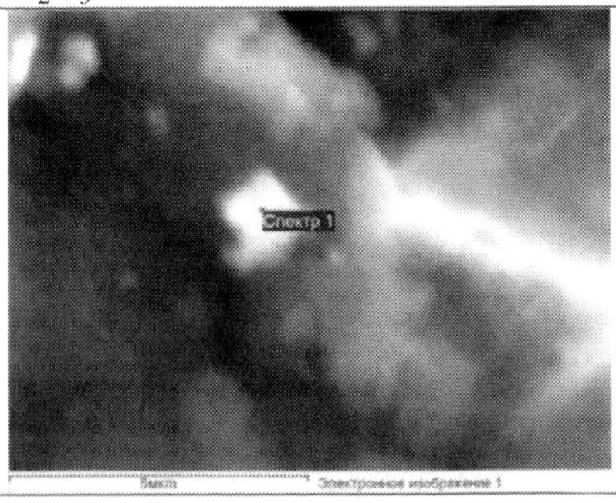

Fig. 3a.

element	gravimetric %	nuclear %
O K	73.05	85.05
Al K	21.77	12.37
Si K	2.59	1.72
Fe K	2.59	0.86
in total	100.00	

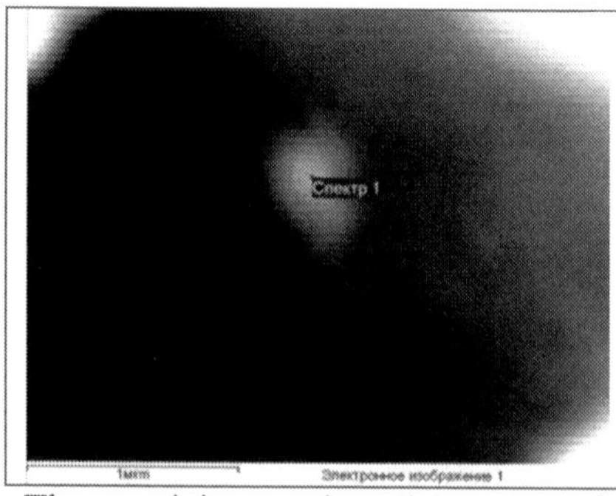

Fig. 3b

element	gravimetric %	nuclear %
O K	54.98	67.93
Al K	37.51	27.48
Si K	5.50	3.87
Fe K	2.01	0.71
in total	100.00	

The remaining sand particles may adversely effect on the adhesion of the ceramic to metal frame because ceramics will bear on the nanoparticle layer of sand Analyzed the influence of aluminum oxide particles on a layer of metal-ceramics. It is known that in the basis of interaction ceramic coating with the metal is based on two factors: molecular and mechanical interactions. In this case, the mechanical action is provided by sandblasting and producing a rough surface. A layer of sand remaining on the surface can result in the formation of microcracks upon mechanical stress and thermal annealing during deposition of ceramics.

Consider the behavior of the system under mechanical load. It is known that the ceramics used in the coating has the following composition: feldspar (a mixture of sodium and potassium), kaolin (a compound aluminum and silicic acid, hydrated aluminum silicate), quartz (silicic acid anhydride) and dyes. Due to elastic deformation in the contact zone transfer contact force during loading may occur at relatively small and changing in size areas of contact, causes considerable stress. The change in specific volume obeys the following rule: $\frac{\Delta V}{V} = \epsilon(1 - 2\mu)$, where ε - relative elongation, μ - Poisson coefficient. For cobalt chromium alloy, this expression has the form ε (1 - 2 * 0,26) = ε * 0,48, for ceramics ε (1 - 2 * 0,22) = ε * 0,56, for the aluminum oxide ε (1 - 2 * 0,32) = ε * 0,38. We make the assumption that the relative elongation in all cases the same, and then the relative change in volume for ceramics in ~ 1.5 times greater than for the aluminum oxide. This in turn will give rise to stresses at the boundary of ceramics and aluminum oxide, and eventually lead to the formation of microcracks. In the process of applying the ceramic can be heated to 1100 ^{0}C. As a result, we are seeing the impact of thermal expansion, namely during cooling due to the difference in thermal expansion coefficients of the ceramic and aluminum oxide voltage will occur at the interface. Thus set possible cause of flaking ceramic from the surface of products.

3. Conclusions

The research resulted in was found that after the sand blasting on the metal surface are sand particles of size 100 - 200 nm are also observed in some places a large particle size of 2 - 3 mm, which is imprinted into the metal. It is shown that these particles can cause crack initiation at the metal-ceramic and lead to her cleavage. It is recommended to find another method of forming a surface microrelief, such as electro physical methods of processing [4].

Acknowledgments

This work was funded by the subsidy allocated to Kazan Federal University for the state assignment in the sphere of scientific activities.

References

[1] Plasma application of protective polymer-powder coatings to ultrasonic sensors. BiomedicalEngineering.2011. T45(5) 138-200

[2] Research of plasma-electrolytic processes in various ratio of the area anode to cathode. Journal of Physics : Conference Series. 479(1), 012005, 2013

[3] Balkenhol M, Ferger P, Mautner MC, Wostmann B. Provisional crown and fixed partial denture materials: mechanical properties and degree of conversion. Dent Mater 2007;23:1574-83

[4] Investigation of the influence of plasma-electrolytic processing on the surface of austenitic chromium-nickel steels.479(1), 012003,2013.

IOP Conf. Series: Materials Science and Engineering **69** (2014) 012018 doi:10.1088/1757-899X/69/1/012018

Engineering and blue-collar occupations prestige in the minds of industrial city students (Naberezhniye Chelny city experience)

A T Kayumov, D N Mustafina
Kazan Federal University,
423800, 68/19 Mira Avenue, Naberezhnye Chelny, Russia

E-mail: chelny@kpfu.ru

Abstract. In this research work an attempt is made to reveal the factors and level of engineering and working professions prestige in the minds of studying youth on the basis of empiric data. The results of research allow revealing the mechanism of social-professional self-identification of school-aged youth, on the basis of which an opportunity to study in a new way traditional system of profession-oriented work appears.

1. Introduction

Economic (market) and political reforms of the last two decades have brought significant changes in the composition, structure, and social stratification of professions, in society values and value orientations of youth and as a consequence in the hierarchy of professions prestige. A large number of scientific studies indicate that, formed in conditions of painful market reforms, hierarchy of occupational prestige is deformed, not meeting the development needs of the society, socio-professional groups, and to a considerable extent, misleading consciousness and behaviour of young people [1; 2; 3]. The prospects of Russian society development are challenged by critically low status and prestige of scientific workers, teachers, doctors, culture workers, engineers, and many working professionals of skilled labor. It negatively influences the views and life plans of youth as well. Thus, problems of professional self-determination of youth, status and prestige of professions are interrelated. In the contemporary circumstances they are not only relevant, but also acquire the priority character of political, economic and moral-ideological problem of the whole society. Issues related to the qualification and creative potential development, quantitative reproduction of engineers and workers primarily on large city-forming enterprises are particularly important for such industrial centres as Naberezhnye Chelny.

2. Main part

For sociology, perspective understanding of the profession is neither a simple separate system function of the social labour division, not single individuals properties but social-group reality. Both professions and to a varying degree institutionalized social groups exist both in the real world and in the symbolic one, in the world of public consciousness and cultural views [4; 5]. In our opinion it is possible to distinguish at least three relatively independent aspects of their lives and activities that should be taken into account in the measurement and analysis of occupational prestige. 1. Assigned functions, role they play in the functioning and development processes of economy and society. For example, the role of scientists and engineers in the technological progress provision, artists and educators in providing the moral, aesthetic and intellectual development of society and the younger generation. According to this aspect there are both retrospective and prospective views in the public consciousness. They are largely mediated by dominant ideologies, political rhetoric, the style and content of media reports. 2. Sustainable socio-cultural features of the group and its typical representatives. They manifest in the level of education, culture and fellowship, in the style of consumption and behaviour, in attributes of clothes, appearance, in the vocabulary and in the speech style. All they differentiate groups and at the same time are of stratigical nature. 3. Third relatively independent aspect is connected with the socio-economic characteristics of the group. They indicate profession socioeconomic status, income, living standards, working conditions, career, realization of needs and interests. In addition to the real and objective existence all these aspects and associated properties are reflected in society's symbolic being.

In methodology of several studies professions are measured at two different yet interrelated properties: the degree of their social recognition and the "profitability" [6; 7].

Public recognition, legitimacy of professions put forth remarkable autonomy from their profitability. According to research materials young people both employed and studying prefers profitable, cost-effective employment forms and fields. For youth social categories and groups engaged in such spheres, as a rule, are referential, embodying life success. In the context of a difficult choice substantive aspects of professions naturally become less important. As we can see paradoxical and marginal types of people in a transitional period probably begin developing, first of all, in the face of today's young generation, typological certain qualities adequate to the market economy model in Russia. We note that the parameter "profession profitability" in our approach is included in the concept of "profession attractiveness" as a leading component. From the standpoint of our approach we cannot agree with the attempts of strict distinction of professional and economic stratification. The economic aspect is the necessary attribute of all professions and distinguishes them from amateur practice. From this point of view close correlations between income indicators of socio-professional groups and professional self-prestige are not random ones.

New concepts such as "image of a profession", "pathos of a profession" need additional analysis to justify their inclusion in concept structure of occupational prestige.

Occupational prestige is considered in terms of their attractiveness for students of a large industrial city. This phenomenon of social and group consciousness manifests on a personal level as relatively stable value relation to the professions. It has multivariate, complicated determination and formation mechanisms; it also executes regulatory and incentive functions [8]. In accordance with our approach attractiveness assessments of occupations reflect first of all socio-economic and stratification characteristics of professional groups.

Naberezhnye Chelny city both as an urban centre and as a complex social environment, where students´ ideas about the occupational prestige are formed, has its typological and "individual" features. This is a relatively young, large, multi-ethnic and fast growing modern city. Gradually losing its monofunctionality and direct dependence on the base of city-forming enterprise, however, it maintains and develops its industrial profile. Transformational crisis of 90s significantly reduced the security, status and prestige of engineers and workers not only of KAMAZ plant, but most of other enterprises of the city. These processes became significantly reflected in the consciousness of the population and especially young people. They contributed to the decrease of the industrial professions´ attractiveness. Large industrial enterprises, as studies show, are significantly in arrears of business, trade, finance and management in the attractiveness of the socio-economic labor conditions [9; 10]. These areas and industries acquire the status of prestigious and promising ones in young people's minds. They become more preferred as spheres of employment, career and objects of professional ambitions. All this leads to a gradual allocation of "youth" industries, on the one hand, and sectors and enterprises with aging contingent workers, on the other one. General demographic decline of the 90s and mass release of the first generation employees of large enterprises out of the economically active age also exacerbate the deficit situation, shortage of personnel. In this difficult context the problem of students´ orientation in engineering and working professions assumes a special relevance. Major factors in this orientation, of course, are the status and prestige characteristics of the most demanded production occupations. Modern youth as the second generation of citizens is the medium of new tendencies more adequate to the requirements and market economy conditions. The empirical base of our research are the data obtained during specific applied sociological research among pupils of the 8th, 9th and 11th forms of secondary schools in Naberezhnye Chelny.

The ten-point scale methodology, proven in the framework of Novosibirsk project, was used to assess the occupational attractiveness within this research. Those surveyed were offered a list of 32 best known, mass, and for some items popular in the city professions. The task of measuring the socio-economic status and characteristics of professions forced us to introduce the concept of "social potential of a profession", which was determined through the list of indicators reflected in the toolkit. More precisely, using this concept the following occupational groups were assessed: workers, engineers and economic ones. Pupils´ value orientation in the sphere of labour and future professional activity were also assessed. They reflect the needs and interests of individuals, and the typical conditions and ways of their implementation. These also reflect stereotypes of mass culture media broadcast, typical characteristics of youth subculture in the transition period, influence of different reference groups. They provide selective attitude to their profession, highlighting the most significant and valuable properties. We proceeded from the assumption that the assessment of occupational prestige is formed in a certain extent on the basis of pupils´ value orientations interaction in the sphere of labour and socio-economic characteristics of professions and their development level. It is about domination in the pupils´ minds of "materialist" values, on the one hand, and attention and interest focus on their socio-economic and, consequently, their status-stratification characteristics, on the other hand. Anticipating the received materials analysis one can say that there is a strong correlation between performance estimates of prestige and social assessments of potential trades.

"Social potential of the profession" is understood, within this research, as the information image of its social opportunities in the consciousness and the submissions of the pupils. Social potential of the profession is a kind of reflection of socially stratified status of professional groups.

Differences between workers and engineering professions in the students' assessment largely reflect real differences in the nature and content of the work of these socio-professional groups and, therefore, can be considered to be functionally justified. However, the differences in the estimates of engineers and technicians occupations on the one hand, and economic professions on the other, are, in our opinion, the reflection of the developed in the last decades deformations and distortions of stratified system of both society and major cities.

Status of economic, "market" professions benefits should be seen as dysfunctional and do not reflect the more complex or more creative work. They are the structural constraints of normal and sustainable development of both the economy and society as a whole. The considered value orientations and evaluations of potential professions underlie notions of prestige and attractiveness of existing in the society of professions and specialties. Of course, they reflect local and regional characteristics of the economy, labour market and employment. In this regard, Naberezhnye Chelny, as well as Tatarstan as a whole differs in sustainable vector of socio-economic development in the last decade. However, this development is still uneven and does not exclude serious imbalances. Sociocultural views of young people, clearly lag behind the movement of the economy and its needs. However, they though with some distortion, but reflect both yesterday's and today's working conditions, characteristics and deformation of socio-professional stratification and at the same time they are both the factors and the development restraints.

In general, the service-sector jobs for students are more attractive than production ones. Consistently the economic and market-based professions are perceived as elite. The study revealed the presence of gender differences in the perception and evaluation of many professions. Professions related to education, culture, health and activity in the service sector tends to be more highly valued by the girls, and manufacturing, technical training preferred by the young men.

3.Conclusion.

According to our hypothesis the youth when assessing the attractiveness of the profession pays their attention mainly to the socio-economic characteristics, but in varying degrees, the socio-cultural features of typical representatives of professional groups are also taken into account. This factor is particularly active when there is a clear mismatch of their appearance and lifestyle normative notions and stereotypes of youth subculture. This is probably one of the manifestations of cognitive and cultural dissonance. Workers today for the most part are not native attributes of success, which is focused by mass media and the youth of today, they differ significantly from the reference group also in lifestyle, consumption and communication. Factors increasing the prestige of working professions in the minds of young people, obviously, are the processes associated not only with the appreciation of the real socio-economic status of their respective groups, but are also associated with changes in the dominant ideology, with the development of a new style of life, communication and self-representation. Much more attention should be paid to the career prospects of young workers in the social policy companies, PR projects, and career-oriented work so that working professions would not be perceived as dead-end branches from the point of view of socio-tier development. The engineering profession according to our study, albeit slowly, but steadily is developing in the direction of improving the social status, potential and prestige. For Naberezhnye Chelny, as for large industrial center, it is a significant fact. This trend, of course, requires political, administrative and information support not only in the framework of a single city or region, but society as a whole. Raising the prestige of the engineering profession as well as working professionals, is a structural, strategic factor in the realization of tasks associated with the new industrialization of the economy with the development of high-tech and information technology and with the transition to innovative model of development.

4. References

[1] Bevzenko L D 2007 *Life success, values and life styles.* // Sociology: theory, methods, marketing (Kiev), – № 4. – Pp. 132-152
[2] Bourdieu P 1994 *Chooses dites.* – M, – 288 pp
[3] Kayumov A T Konikov F K 2007 *Value orientation of pedagogical higher school graduates //* Science and school. – №1. – Pp.56-59
[4]. Shkaratan O I 2009 *Socio-economic inequality and its reproduction in modern Russia.* - M: CJSC "OLMA Media Group"– 560 pp
[5]Shubkin V N 2010 *Sociology and society: Scientific knowledge and science ethics.* Monograph. – M., – 424 pp
[6] Gorshkov M K Shereguy F E 2010 *Youth of Russia: a sociological portrait.* – M., 2010. – 592 pp.
[7] *Revolution claims and changes in life orientations of the youth*: 1985-1995. Ed. V.S. Maguna. - M: Institute of sociology RAS – 312 pp
[8] Cherednichenko G A 2005 *Personal plans of secondary school graduates //* Sociological studies. – №7. – Pp.114-117
[9] Semenov V E 2007 *Value orientations of modern youth //* Sociological studies. – № 4. – Pp.37-43
[10] Chuprov V I 1998 *Young people in social reproduction //* Sociological studies. – №3. – Pp.93-99

IOP Conf. Series: Materials Science and Engineering **69** (2014) 012019 doi:10.1088/1757-899X/69/1/012019

Surface hardening of components by automatic plasma electrothermal installation with molten cathode

A A Khafizov, B Yu Shakirov, Yu I Shakirov
[1]Senior Lecturer, *Kazan Federal University*
423800, Republic of Tatarstan, Naberezhnye Chelny, 68/19 Mira avenue

E-mail: chelny@kpfu.ru

Abstract. An automatic installation for surface hardening of components has been designed. It enables to enhance the effectiveness of technological process and to provide the stability of quality indexes.

1. Introduction

The success achieved in the field of surface hardening of materials by plasma current shows high efficiency of this method and requires further development.

Automating the process of heat treatment of components in mechanical engineering can improve the efficiency of technological process (TP). To obtain the desired mechanical properties of metal surfaces modern technologies of hardening are applied [1; 2; 3; 4; 5]. Plasma methods of wear resistant coating or physico-chemical modification of the surface layer of components can reduce the duration of TP, power consumption, save the basic geometric dimensions of components [6; 7, 8, 9].

Problem statement. Ensuring the stability of quality indexes of component a surface, as well as strong economic performance requires the use of TP in automated control systems. On this basis, a promising way to improve technological plasma electro thermal installation (TPETI) for components heat treatment is development of the head of TPETI, automatic control systems (ACS).

2. Basic part. The electrical characteristics of the discharge generated between a hard anode and molten cathode have been researched. Experimental investigations of electric discharge between a metal anode and electrolytic cathode at atmospheric pressure were held in a wide range of current I=0.5-10A, voltage U=0.5 kV, interelectrode distance l=5-50 mm, the diameter of the anode da=5-40 mm for various composition (technical and purified water, solutions of $CuSO_4$, NaCl in technical water) and electrolyte concentration. In experiments, as a metal anode, steel of various grades (Steel 3, St35, St45) and copper of different geometric form (solid cylinder, hollow cylinder, hollow truncated cone, solid cone) were used.

The most important parameter of the electric discharge is the value of density of a current on the anode [10; 11]. As it can be seen from Fig. 1, with the increase of the discharge current up to I=1A a decline of current density can be observed. There is an growth of the spot on the surface of a metal anode. At the critical value of I=1A is the spot covers the lower part of the cylindrical electrode. At the further increase of the discharge current the area of the anode spot is not changed. This fact explains the growth of density of the current on a metal anode with the increase of the discharge current. The change of the interelectrode distance does not affect the nature of the density of a current. (Figure 1, curves 1 and 2). Comparing the curves 1 and 2, you can see that the increase of interelectrode distance leads to some reduction of current density.

IOP Conf. Series: Materials Science and Engineering **69** (2014) 012019 doi:10.1088/1757-899X/69/1/012019

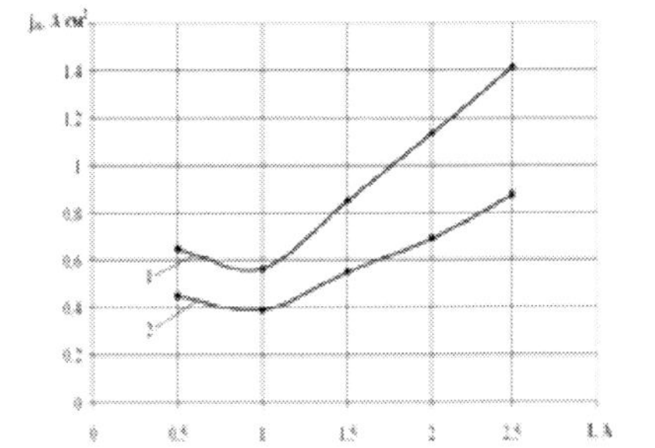

Fig. 1 - Dependence of current density on the metal anode from the current of the discharge for different interelectrode distances: 1 - l=5 mm; 2 - l=15 mm

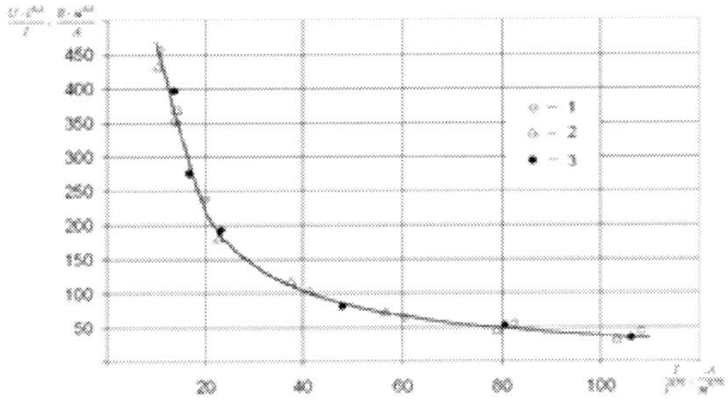

Fig. 2 - The generalized volt-ampere characteristic of the discharge between electrolytic cathode of technical water and metal anode at da=15 mm: 1 - l=3 mm, 2 - l=7 mm; 3 - l=13 mm

With the help of Excel and MathCad software for discharge between electrolytic cathode of technical water and metal anode the generalized criteria equation in the following form has been worked:

$$\frac{U \cdot l^{0,5}}{I} = 5752\left(\frac{I}{l^{0,75}}\right)^{-1,1}.$$

The dependence of $\frac{U \cdot l^{0,5}}{I}$ from the parameter $\frac{I}{l^{0,75}}$ represented in Fig. 2.

The maximum standard deviation of the experimental values of voltage discharge from the obtained equations is less than 10%, so these criteria equations can be used in the calculations of plasma installations with electrolytic cathode. To solve the set tasks the head of TPETI has been developed (Fig.3). It is made of quartz glass, which is water-cooled. The plasma gas, which is shown by arrows, is delivered to the channel. The discharge occurs between the cooled copper anode and electrolyte.

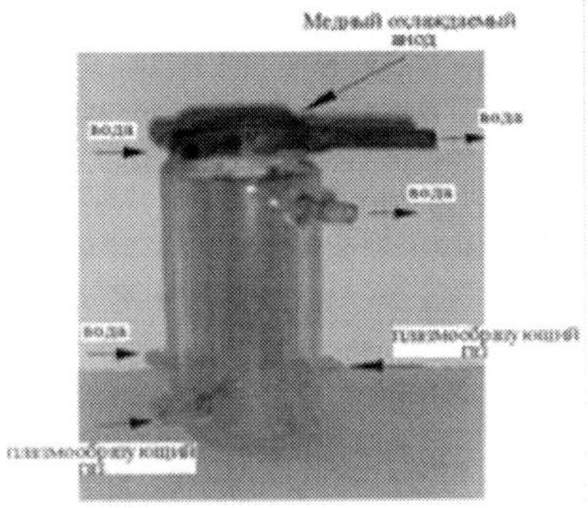

Fig. 3 The picture of the head of technological plasma electrothermal installation.

Technological process of plasma thermo strengthening can be used in the processing of outward of cylindrical surfaces for such details as a worm, a roller, a windrow.
The efficiency of the process of plasma thermo strengthening depends on the thermo physical properties of the processed metal and technological factors [12]: the energy potential of the plasma jet, rotation speed of the component, the gap between the part and the head of TPETI, the intensity of water cooling and the temperature of the plasma jet (approximately 10000°C).

Hollow cylindrical components made of steel 45 have undergone plasma processing . Metallographic and durometric analyses were performed on the samples in the form of plates cut from the component perpendicular to the plasma process.

In this case the results of studying the structure and microhardness edgewise and in depth of the hardened layer were the criteria of assessment hardening.

It is shown that the character of changes in the microhardness in the depth of the hardened zone virtually identical for different modes of plasma processing characterized by the presence of three basic sections.

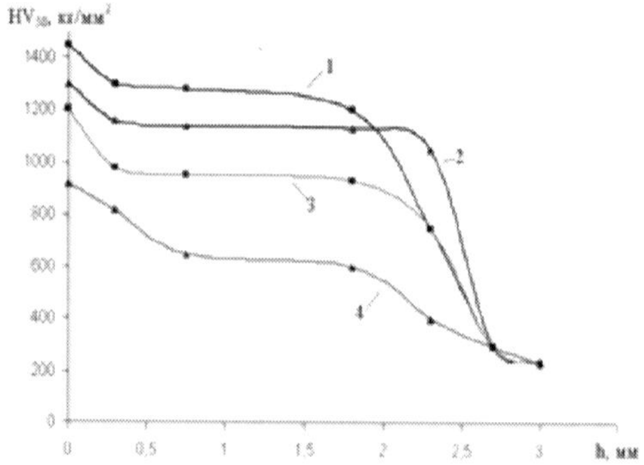

Fig.4. The character of the distribution of microhardness in the depth of the thermo hardened layer depending on the modes of plasma processing: 1.2 - change of the gap between the cylinder and TPETI and the component; 3.4 - changes in electric current intensity.

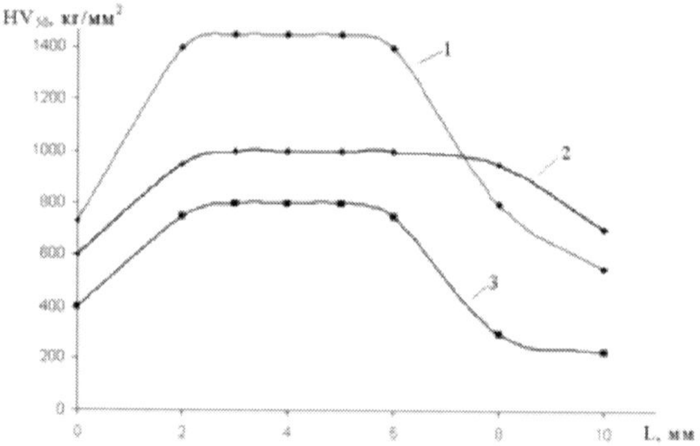

Fig.5. The character of distribution of microhardness edgewise of the zone of heat treatment depending on the electric current intensity.

3. Conclusion. The developed head and control system ensures the stability of parameters of surface hardening components quality. For optimal hardening of working surfaces of cylindrical components, you need to choose the modes of processing, which would ensure the minimum width and depth of the transition zone. In addition, the necessary condition for high-quality processing is the stability of parameters of heat-treated plots microhardness, which provides high service properties of the processed surface.

4.References

[1]. Romankiv P G 1975 *Mass transfer processes of chemical technology (systems with solid phase)* Leningrad: Khimiya , 336

[2]. Galyautdinov R T, Kasparov N F, Luchkin G S 2003 *Inzhenerno-Fizicheskii Zhurnal* **75 (5)** 170-173

[3] . Ahunov DN 2010 *Software defined physicochemical properties of polymeric compositions in the manufacture of cellulose nitrate* - Electronic Journal "Education and Science Zakamja Tatarstan" , Naberezhnye Chelny , INECA, № 18 , http://www.nauctat.ru.
[4] . Deacons S G 2010 *Optimal design of industrial devices of chemical technologies based on physical and mathematical modeling* Herald Kasane . tehnol . Univ . № 4 131-140
[5] . Makhmutova D I 2011*Model regeneration process technology components of energetic condensed systems* Herald Kasane . tehnol . Univ . № 13.156 -160
[6] . Akselrud GA 1974 *Extraction (system solid - liquid)* Leningrad: Khimiya. 256
[7]. Romankiv P G 1983 *Extraction of solids* Leningrad: Khimiya , . - 256
[8]. Kurochkin M I 1978 *The extraction and solids leaching . (Library of work - Chemist)* Leningrad: Khimiya , 72
[9]. Lykov A V 1978 *Heat and mass transfer : Reference* Moscow: Energiya , 480
[10]. Rudobashta S P 1980 *Mass transport systems in the solid phase* Moscow: Khimiya 248
[11] . Tikhonov A P 1976 *Equations of Mathematical Physics* Moscow: Nauka 724
[12]. Samara A A 2001 *Mathematical modeling . Ideas. Methods . Examples* 2nd ed . - Fizmatlit, Moscow.
[13]. Karpova M N 2011 *Mathematical model of mass transfer in polymer products* Herald Kasane . tehnol . Univ. № 21. - With . 215-218 .

Competitiveness assessment of engineering products

A R Kharisova, A S Puryaev
Kazan Federal University,
423800, 68/19 Mira avenue, Naberezhnye Chelny, Russia

E-mail: chelny@kpfu.ru

Abstract. This article discusses the problem of increasing the competitiveness of the engineering industry through the implementation of innovative projects. Based on the analysis of the features of innovative projects formulated a conclusion according to which the innovative projects effectiveness evaluation should take into account non-economic indicators such as social, ecological, resource, scientific and technological. We formulate the process and provide a methodology to evaluate the effectiveness of innovative projects based on non-economic indicators. This technique is aimed at assessing the projects increase the competitiveness of products, which is understood as a comprehensive line of products a whole range of different physical limitations of the essence, allowing the long run to get sustainable income.

1. Introduction

One of the most important aspects of economic development is the growth of production in all sectors and at every single enterprise. In conditions of a market economy and in connection with Russia's accession in WTO the important thing is increasing of business and industry competitiveness in general. The problem of nowadays competitiveness is very serious, particularly in engineering. Russia has long been staying as a leader in the field of engineering, but recently the powerful production of developing countries makes strong competition of domestic products. On the background of growing industries and increased competition special importance acquire innovation activities of enterprises, in particular the development and implementation of innovative projects. Effectiveness of innovative projects mostly depends on its proper evaluation, which gives an opportunity to determine the best alternative project that satisfies a whole range of restrictions of different physics. It's obvious that any innovation must be new, practically feasible, must bring economic and other benefits. For the process of innovation need a project to be effective [1]. Innovative project is the statement of reasons for the promotion of novelty (innovation in particular) and the process of novelty, by the means of investing [2; 3]. Practicability of investing funds for the development of innovation is necessary to be justified so that innovative project met certain criteria. Set of criteria depends on the specifics of the engineering industry and businesses in particular. The main criterion is the economic practicability, expressed through a system of economic indicators. For evaluating the effectiveness of innovative projects, as a rule the method of «Cash flow» is used.

2. The basic part

«Cash flow» method is only able to evaluate the economical cost-effectiveness of the innovation project. M.A. Bendikov said: "the main criteria for the effectiveness of the innovative project is the assessment of its technical and economic feasibility in terms of enterprise, identification and evaluation of its effect on the region, its economy, the environment, etc. " [4] . I.e. effectiveness of the innovation project can't be considered only from the point of view of economic efficiency [5; 6]. Certainly, economic efficiency is decisive in choosing the innovative project, however, in our opinion, *non-economic performance indicators* should also be considered, such as social, environmental, scientific and technical effects, etc., as they affect the competitiveness of the products, companies, industries and countries. Cost-effectiveness is not the same as social and is inherently "blocker" of socially useful innovation [7; 8]. Existing methodologies of innovation project evaluation often based on quantitative assessments and take into account only the economic benefit from its implementation [9].

According to our point of view, it's rational to divide the process of evaluation process of innovation project on two stages [10]:

Estimation of efficiency of an innovative project based on non-economic indicators;

Estimation of efficiency (in particular of the feasibility of an innovative project).

Such a division is dictated by the fact that for innovative projects that are implemented by investors as a State or State funds, particularly important meaning get projects, beneficial in the terms of improving the environment, technology and the image of the state, that's why the evaluation of economic efficiency becomes a secondary task.

At the first stage innovative projects using non-economic indicators are estimated. Since these indicators have mainly a high-grade basis, transfer them into a comparable form, i.e. in dimensionless units. With the help of expert assessments determine the coefficient of importance of each indicator. Besides the expert estimates, is also applicable Fishburne formula for cases where indicators can be ranked only by degree of their importance: "not significant", "significant", etc.:

$$\alpha_i = \frac{(n-i+1)}{n(n+1)} \quad (1)$$

where α_i – index of indicators' importance of i category; n – is the number of categories of innovative projects' performance indicators, i – number of the appropriate category.

Then analyze evaluation of each indicator by the following formulas:

$$s_{ij} = \frac{SF_{ij}}{S\max_{ij}} \quad (2)$$

where SF_{ij} – actual index value i for project j; $S_{\max ij}$ – maximum possible index value indicator of i for project j, if needed growth i indicator.

$$s_{ij} = \frac{S\min_{ij}}{SF_{ij}} \quad (3)$$

where SF_{ij} – actual index value of indicator i for project j; $S\min_{ij}$ – minimum possible index value indicator of i for project j, if reducing of indicator i meaning is needed.

Thus, synthetic (Final) indicator of project is defined by the formula:

$$S_j = \sum_{i=1}^{n} \alpha_i s_{ij} \quad (4)$$

where S_j – aggregate data of project j; α_i – index of importance (weight) of indicators i category; s_{ij} – evaluations of indicator i of project j.

The meanings of aggregate data are in the range from 0 to 1. The closer the meaning of aggregate data to 1, the more preferred project, i.e. such a project is the best for improving of the competitiveness of engineering products, as considers the limitations of other physical entities (environmental, social, ergonomic, technical-technological etc.).

In the second stage using methods of cash flow the cost-effectiveness of projects is defined (counting indicators NPV, IRR, PI, DPP). We choose the one among effective from an economic point of view the projects that has the maximum yield.

The method described above should be used to evaluate innovative projects aimed at improving the competitiveness of engineering products. Thus, under competiveness the complex product accordance to a whole range of restrictions of different physics is meant, allowing to get a steady income in the long term, not just the common concept of value approach: engineering products, bringing profit is competitive. I.e. technique aimed at projects where investors and company's management team have set themselves the task not only in generating profits from the project, but also tasks related to the improvement of environmental, social, scientific and technological situation in the enterprise of the region and the country, which will undoubtedly improve the company's competitiveness in the domestic and international markets in the long term and for the long time lag.

3. References

[1] Sorokin N T 2009 *Problems and trends of the Russian engineering* // Automotive industry № 1
[2] Ribkina E A, Puryaev A S 2006 *System of innovation attractiveness in the engineering enterprise* (on example of "RIAT" Naberezhnye Chelny) // «Mechanical engineer» №3 P 11–17
[3] Ribkina E A, Puryaev A S 2006 *Evaluation and improvement of the innovation attractiveness system in the engineering enterprise appeal* // «Mechanical engineer» №4 P 14–18
[4] Bendikov M A 2001 Realizability assessment of the innovative project // Management in Russia and abroad №2 P 37-43
[5] Tumina T A 2009 *Assessment methodology of the effectiveness of innovation* // Transportation business in Russia №1 P 31-42
[6] Hozhaev I S 2010 *Improvement of methods for evaluating the effectiveness of innovative projects*: Autoref. of dis. Candidate of Economic Sciences Belgorod
[7] Sukharev O S 2011 *Economic development and modernization task* / / Horizons of economy №3 P 19-29
[8] Yashin S N, Boronin O S 2011 *Theoretical approaches to evaluating the effectiveness of innovative projects* / / Proceedings of the Nizhny Novgorod State University by the name of R.E. Alekseyev, Nizhny Novgorod №2 P 248-254
[9] Nizamova A Sh 2012 *Estimation of efficiency of innovative projects by indistinct sets* // Management of economic systems: an electronic journal №4 (40) URL: http://www.uecs.ru/uecs40-402012/item/1241-2012-04-10-06-36-09 (date of handling: 01.05.2013)
[10] Kharisova A R 2012 *Problems of assessing the effectiveness of innovative projects* // Science and Innovation: Proceedings of VIII International Scientific and Practical Conference, Prezmyshl P 39-43

Automation System Goals for the Creation and Operation of the Tool

R M Khisamutdinov
Chief technologist of JSC "KAMAZ", Director of Technological centre

E-mail: Ravil.Hisamutdinov@kamaz.org

M R Khisamutdinov
Kazan Federal University,
423800, 68/19 Mira avenue, Naberezhnye Chelny, Russia

E-mail: chelny@kpfu.ru

Abstract. Complex automation of processes for creation and operating the instrument, consistent linking of hierarchical levels in the single system of collection and processing of data and operations management, integration with TEAMCENTRE PLM and SAP/R3 ERP significantly improve the quality and efficiency of production preparation.

1. Introduction

Implementation of the automated system for the creation and operation of the instrument (ASCOI) of engineering company with intellectual support of the automated system for the selection and design tool for machining of complex product surface and automated control of the production preparation in the process of establishing and operating the tool with their integration into a single information system of the enterprise is an important scientific production problem. [1,2]

For the formation of intelligent automated systems for the selection and design tool, it is necessary to develop its structural-hierarchical model, which would reflect a modern state of scientific knowledge in this field.

2. Basic part

The need to develop invariant structural-hierarchical model of the instrument is connected with the formation of the intellectual support of the automated unit for the selection and design of the instrument as an integral part of ASCOI enterprises. The proposed model is flexible and allows you to embed any structural unit as an indication of the instrument.

The block of automated intellectual support for establishing and forming tool built on block-modular principle. The unit includes a database tool (database), the module of computer simulation of surface treatment of components, module profiling forming tooling surfaces (TS), the module the design of the structural parameters of the tool, as well as means of computer graphics. [5,6]

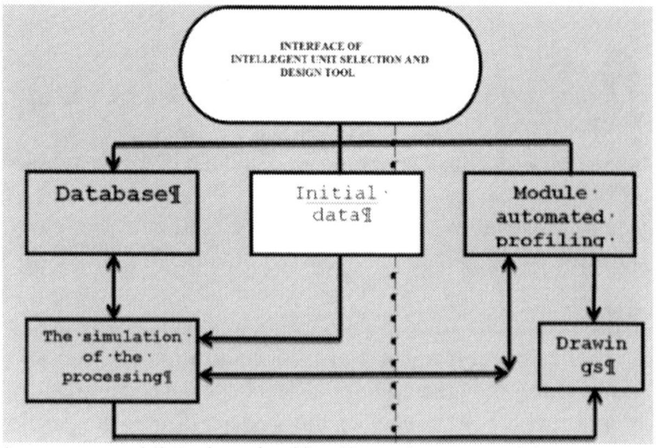

Fig. 1. Enlarged structure and links in unit of intellectual support of the automated selection and forming tool.

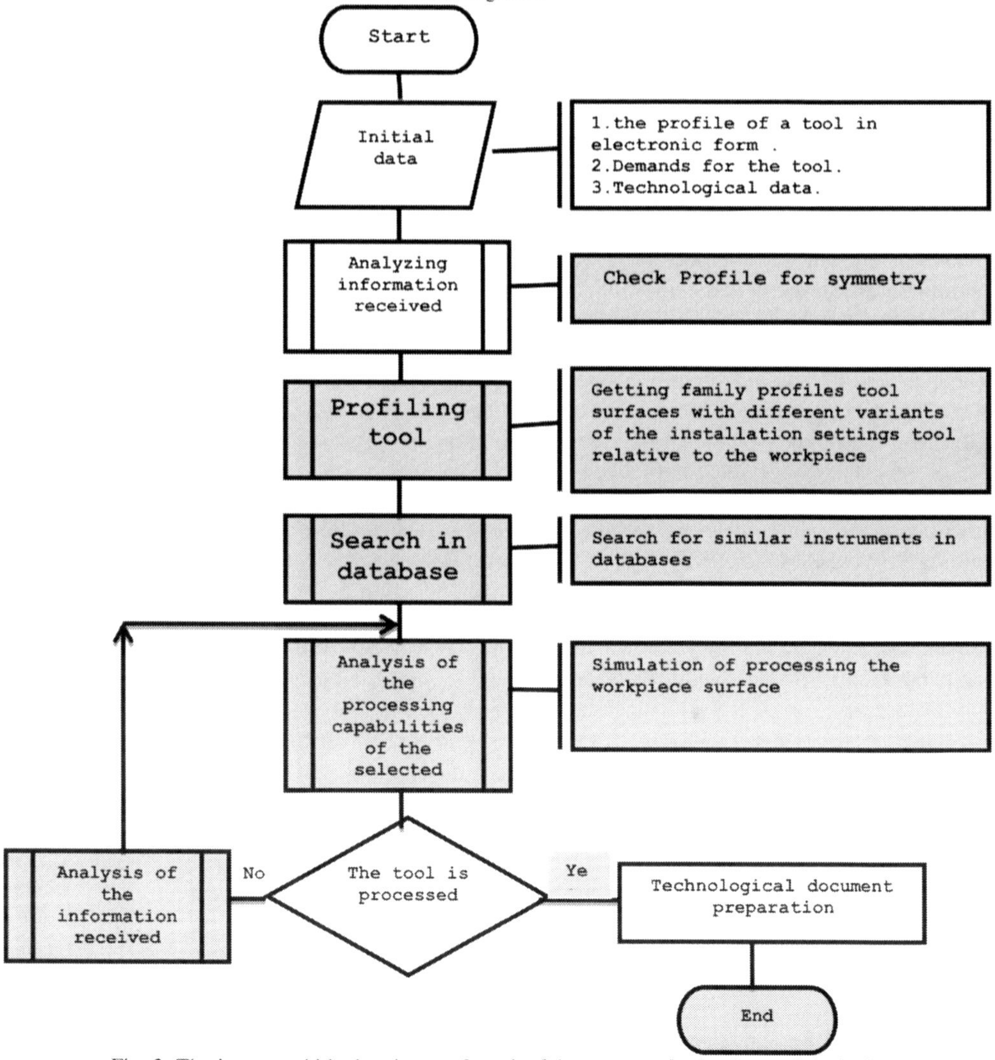

Fig. 2. The integrated block-scheme of work of the system of selection and designing tool.

The calculation of results are passed to the DB tool to pick a similar tool. An important task is to identify the profile of the instrument stored in the database. [7] Comparison of theoretical and profiles stored in the database is performed by the algorithm presented in figure 2.

If the solution using information from the database did not lead to the desired solution, we introduce a special stage for forming systems - design module design parameters of the tool, which is closely linked with the system forming TS, computer simulation system and database.[8]

Initial data for design module of design parameters are geometric parameters of the tools, previously calculated profile forming a TS, and information about types of basic material (BM) and instrumental material (IM), the technological process of manufacturing of details, passport characteristics of the components of technological system. Since the profile forming TS is defined when performing a selection tool, its design is the calculation of the connecting part on the known correlations and making a drawing. [8]

To create a database tool preceded by the analysis of surfaces of details and the link between the basic surfaces of parts and tools, which can process them.

In practice, the actual effect of scientific methods with application of means of automation is attained only when various systems integrated into a single information space, only this solution gives a general synergetic effect, measured in terms of improving quality, the maximum human factor in the reduction of deadlines and cost optimization. That is why there is an acute need in development of the automated control system of manufacturing preparation (ACSMP), due to the fact that the major machine-building companies, where there is a simultaneous preparation of production on thousands of original parts and tens of thousands of positions of the developed tool, it is very difficult to ensure the functioning of the developed system through design, production and implementation of the tool. [9]

The review of the market of published software revealed that a system that meets these requirements does not exist.

The package of applied programs intended for the effective management of production large machine-building enterprise in the process of establishing and operating the tool has been developed.

Thus, the developed software is a complete solution for organization of work on achieving the targets of the enterprise production of a new product. [10]

Introduction of the developed ACSMP helped to ensure the design, manufacture prototypes of the original instrument and conducting a trial of treatment in 4 times faster than the average performance of holding TPP in KAMAZ OJSC.

Complex automation of processes of creation and operating the tool to achieve the required geometric precision and quality of products with complex surfaces, as well as automated and intelligent selection, design and control instrument taking into account the laws and the relations between the geometric parameters of the details and geometrical parameters of the tool, consistent linking of hierarchical levels in the single system of collection and processing of data and operations management, integration with TEAMCENTRE PLM and SAP/R3 ERP allow to raise quality and the efficiency of production preparation considerably.

3.References

[1] Simonova L A, Khisamutdinov M R *Integration of local information systems with the SAP R/3.* 2012 Kama readings: interregional scientific-practical conference, G. N. Chelny, pp. 155-159.
[2] Simonova L A, Khisamutdinov M R *The problem of the exchange of data between ERP and PLM systems in machine building industry. New technologies of science-intensive machine-building: development priorities and training: an international scientific-practical conference,* 2013 G. N. Chelny, pp. 341-345.
[3] Khairullin A H, Simonov L A. Khisamutdinov R M, *Some principles of modeling and algorithm of management processes instrumentalisation for machine-building enterprises* 2011 KSPD №4. - pp. 32- 38. 13.

[4] Khisamutdinov R M, Khisamutdinov M R *New technology of production preparation on the basis of models, methods and means of automation of processes of creation and operation of the instrument* 2012 Theses of reports of International scientific-practical conference "Education, science and production. New technologies as a tool of strategy of development and modernization 2020. -Kazan, June 19, 2012, Kazan: Publishing house "MSPRESS" p.276.

[5] Khisamutdinov R M, *Strategy equipment of technological processes of the enterprises of group KAMAZ" in modern economic conditions* 2008 Proceedings and abstracts of VII Congress of technologists of automotive industry 28 February Moscow, IEC "Crocus Expo". - pp.17-26.

[6] Khisamutdinov R M *Automation of solutions search for the design tool based on modeling of the process and the information - reference system* 2011 Forging and stamping production processing of materials by pressure. №5. - S.43-48.

[7] Adgamov R I, Dmitriev S V, Khisamutdinov R M *Strategy of providing a tool plants of JSC "KAMAZ" in modern economic conditions* 2007 Inter-industry almanac. Business glory of Russia. №4-5.-pp.182-186.

[8] Khisamutdinov R M *Formation of functional connections between the parameters of parts and tools for automated systems of technological preparation of production* 2011 Press - forging manufacture of materials by pressure. № pp. 38-43.

[9] Simonova L A, Khairullin A H, Minnakhmetov I R *The construction of an intelligent management model of the distributed information systems on the basis of genetic algorithms,* 2009 Press- Forging production. Processing of metals by pressure, №8, Page 16-22.

[10] Nurul Nuha Abdul Molok. *The information systems integration plan,* Saarbruecken, "VDM Verlag", 2010.

Innovative Mechanical Engineering Technologies, Equipment and Materials-2013 IOP Publishing
IOP Conf. Series: Materials Science and Engineering **69** (2014) 012022 doi:10.1088/1757-899X/69/1/012022

Diagnosis of CNC machine tools in terms of circular interpolation's accuracy figure

R M Khusainov, S F Belov, O V Chukhontseva
Kazan Federal University,
423800, 68/19 Mira avenue, Naberezhnye Chelny, Russia

E-mail: chelny@kpfu.ru

Abstract. The article deals with circular interpolation test on CNC machines and determination on the basis of their performance the machine accuracy.

1. Introduction

Accuracy testing of circular path is a comprehensive assessment of the geometric accuracy of machine tools with CNC. Almost all factors influencing the geometrical accuracy parameters and therefore rejecting a mobile unit motion trajectory conduce to the fact that the radius of the actual work piece circumference deviates from the programmed circle one. Hence, the reverse is also true - if measured coordinates of the actual circular path and compare them with the programmed one, applying appropriate calculation methods, one can obtain the deviations figures of the machine accuracy, and overall technical assessment [1; 2; 3].

2. Basic part

Standard [4; 5] provides performing this type tests of reference mandrel spinning. But it does not set any specific testing methods, neither the methodology of processing results. One can propose the following procedure for conducting tests (Fig. 1, Fig. 2):

1. Install and secure control frame 3 in the spindle. Fit on the holder 2 on the machine table.
2. Combine the axis of the holder 2 and mandrel 3 (Fig. 1) for example by using an edge finder. Navigate in the CNC system to the control field «Options», into the category of «Zero shift». Identify the actual coordinates X and Y as zero program, for example under the address G54.
3. Move the machine table at the level of about 50 mm in the X-axis with manual data entry (MDI).
4. Move the spindle stock down so that the arbor 3 fit the bar notch (Fig. 2).
5. Install and secure indicator 1 in the strap, set it to «0».
6. In MDI mode enter and practice circular interpolation program clockwise (*G2*) recording indicator in regular trajectory intervals.
7. Graph deviations δr depending on the angle θ in polar coordinates:

IOP Conf. Series: Materials Science and Engineering **69** (2014) 012022 doi:10.1088/1757-899X/69/1/012022

Figure 1

Figure 2

$\delta r = f(\theta)$ 1).

8. Similarly to paragraph 6 work out the circular interpolation program counterclockwise (*G3*). Plot in the same coordinate system, which in paragraph 7.

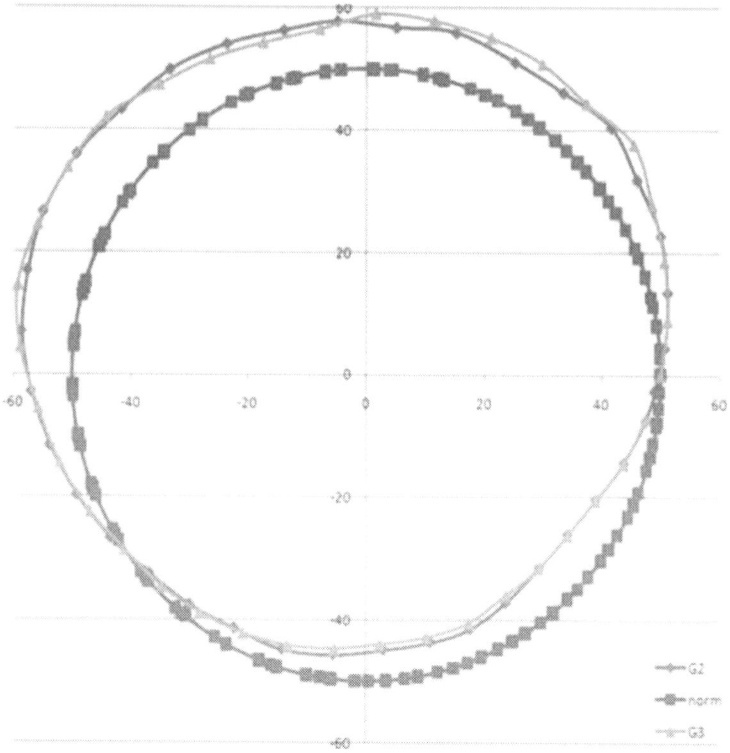

Figure 3 - Graphics-precision testing circular path.

G2 - actual trajectory when moving clockwise, G3 - actual trajectory when moving counterclockwise, norm - a perfect circle. Magnitudes in the graph are in microns.

According to the results one can define the following geometrical errors of the machine:

1. The circular interpolation deviation from circular trajectory. The highest magnitude δr is accepted as a result .

2. The axis positioning errors [6]. Considering the small increment of the polar angle θ_j, at each step, one can define the axis positioning errors, for example Y axis, as δr projections on the Y axis in a positive direction:

$$\Delta Y_j \uparrow = r(\theta_j)\sin\theta_j; \quad \theta_j \in \lfloor 0;180^0 \lfloor \qquad 2).$$

determines The positioning errors are determined in the interval from 180^0 until 0^0 in the negative direction:

$$\Delta Y_j \downarrow = r(\theta_j)\sin\theta_j; \quad \theta_j \in \lfloor 180^0;0 \lfloor \qquad 3).$$

The calculated data can be used to determine the standard indicators of the positioning accuracy [6] and to build the chart (Fig. 4).

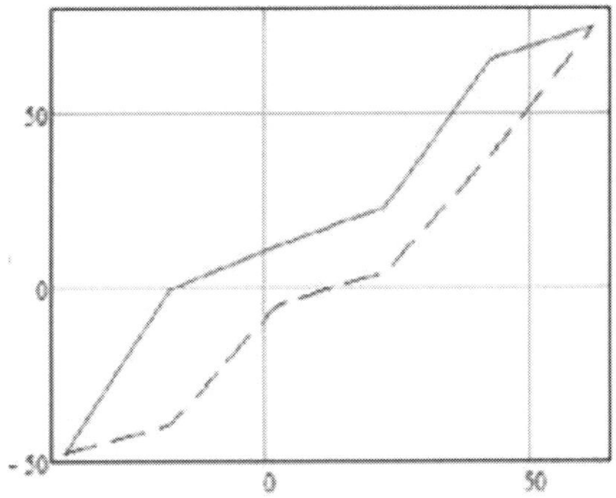

Figure 4 – Graphs of positioning error on the Y axis. The X-axis is the displacement magnitude, mm; the ordinate axis – error, mcm

The indicators of positioning accuracy in the X-axis direction are defined by the same token.

3. The backlashes in the actuator supply.

The testing showed that the backlash is determined as the difference of coordinates of the positioning error graphs on the same length of travel. Processing the obtained data, one can find the maximum and average magnitudes of the backlash on the axis. The average one can be recorded in the memory system CNC for the software error compensation.

On the basis of diagnostic results, one can:

•assess the machine capabilities to ensure the necessary machining precision;

•assess the ability of software errors compensation [3];

•determine the necessity and concrete measures to eliminate the causes of errors by adjustment or repair [7; 8];

•assign a works schedule on maintenance service and repair of the machine with identified errors, technological and economic characteristics of concrete production [9]. The accurate identifying the

IOP Conf. Series: Materials Science and Engineering **69** (2014) 012022 doi:10.1088/1757-899X/69/1/012022

exact faults of a machine allows to organize an effective and targeted service, thereby reducing the machine equipment downtime.

3.References

[1]. Sinopalnikov VA, Grigoriev S.N., *Reliability and diagnostics of technological systems: Textbook* 2005 M.: High school, 343 pp.

[2]. A G Skhirtladze, Ukolov MS, Skvortsov AV *Reliability and diagnostics of technological systems: Textbook* 2008 Moscow: Novoe Znanie, 518 pp.

[3]. VV Yurkevich *Control and diagnostics in automated production. Mashinostroitel*, 2003, №1 - №5.

[4]. GOST 30544-97. *Metal-cutting machines. Methods of verifying the accuracy and consistency of circular path testing*. 2001 - M: Publishing house of standards, 6 pp.

[5]. *Russian National Standard ISO 230-1* - 2010. Testing machines. Part 1. Methods of geometrical parameters measurement. M, 2011 - 90 pp.

[6]. *GOST 27843-2006. Testing machines. The definition of accuracy and repeatability of positioning axes with numerical control*. - M: Standartinform, 2007 - 7 pp.

[7]. *The model time norms for maintenance of machine tools with numerical program control and robots (manipulators)*. M: Economica, 1990, 64 pp.

[8]. R M Khusainov *Reliability of technological systems. "Design and analysis of technical systems: the interuniversity collection of scientific works*. 2012 Issue №4 (18)/ edited by V.G. Shibakov – Nab. Chelny: Publishing house of Kama State Academy of engineering and economics

[9]. A I Yaschura *System of technical service and repair of industrial equipment: Handbook*. M: Publishing house SC ANAS, 2006, 360 pp.

[1]0. Siemens SINUMERIK 840Dsl/828D. *Extended programming. A programming Handbook*. 2011 6FC5398-2BP40-2PA0. Siemens AG. 09/2011 – 609 pp.

IOP Conf. Series: Materials Science and Engineering **69** (2014) 012023 doi:10.1088/1757-899X/69/1/012023

Development of the intellectual system for predicting the properties of compacted graphite iron

L A Simonova, K V Klochkova
Kazan Federal University,
423800, 68/19 Mira avenue, Naberezhnye Chelny, Russia

E-mail: chelny@kpfu.ru

Abstract. The paper considers the ways of reduction of the metal consumption and the cost of castings and improving their physical and mechanical properties and operational characteristics in iron foundries.

1.Introduction

Technical progress in iron foundries aims to reduce the metal consumption and the cost of castings and to improve their physical and mechanical properties and operational characteristics. In this context, a perspective kind of compacted graphite iron with the so-called vermicular (worm-like) shape of graphite is of a particular interest. Currently it is the most popular and promising material in the manufacturing of parts in mechanical engineering (clutch housing, clutch divider transmission, cylinder block, transmission housing). The perspectives of application of compacted graphite iron (CGI) in mechanical engineering are caused by some economic, technological and operational advantages in comparison with conventional casting alloys. Currently in foundry manufacture, particularly in the production of cast iron, the properties of the final products are determined by the control samples obtained from the relevant series of the batch of parts. There are no methods that allow to predict the properties of the final articles based on the information about the components. As compacted graphite iron is modified iron, the selection of the optimal composition and characterization of CGI requires a large number of experimental studies, which causes high costs in production. In this connection, there was a need to create the systems using the elements of artificial intelligence to predict the properties of CGI on the basis of information about the components with the possibility of adaption and self- learning of the system, if we change the input parameters and evaluation criteria. As a result, it will allow to achieve the reduction of laboriousness and the number of experimental studies. It will, undoubtedly, impact on improving the quality and reducing the cost of manufactured products.

2.Shortcomings of existing CAD

In the analysis of existing CAD systems a number of shortcomings is revealed:

1. high cost of CAD;
2. high costs of CAD data maintenance;
3. highly qualified staff;
4. the need to purchase additional modules for modeling different kinds of casting;
5. complexity of modernization;

6. impossibility (in many of them) of optimization of the composition for a given value of the parameter;
7. high demands on the machine resources.

The proposed solution is in the use of elements of artificial intelligence, self-learning neural networks and combination of mathematical analysis methods that will increase the speed of system response on the change of the initial data. Application of the popular software will reduce the cost and simplify the requirements for machine resources.

3.The stages of intellectual system

The system design is divided into two stages: the training stage and the designing stage. The main aim of the training stage is to train the neural network using genetic algorithm. Then, the obtained results are compared with the results contained in the knowledge base. If such data are not available, the experiments are conducted. The results obtained in the experiments are compared with the results obtained by using the neural network. If the results match, they are recorded in the knowledge base. If the results are not confirmed, the initial data are corrected by the expert and the whole cycle repeats.

The process of conducting experiments is divided into several stages. Preliminarily the planning of the experiments by the method of the combinational squares is carried out, that will reduce their number in n^{m-2} times (m - number of factors, n- number of combinations of factors). After the experiment, statistical processing of the experimental data is conducted by the method of combinational squares, to determine the impact degree of each factor on the final properties of the material, and correlation and regression analysis is carried out to find out the dependency of elements from each other. The data obtained are recorded in the knowledge base.

After the neural network is trained, you can start to work with the system. Chemical elements and final product are going to be the input parameters. Based upon the data about the final form of the product it is possible to determine the wall thickness of the article, as the speed and time of solidification of cast iron strictly depends on it.

The stage of work is divided into several blocks.

The first block is the block of initial data formation. Then the program accesses the knowledge base, which includes a database, a base of precedents, rules bases, a block of the image-making formation and a block of searching the similar images with the help of the base of precedent. The data of the base of precedents are used to search for data options with similar characteristics. Precedent is an array of data that describe existing facilities and processes. To make the base of precedents work, it is necessary to work out the rules that help to search the precedent. If the precise values that are specified by the user are not found, then the values in the range are + _3 % from specified are searched. If the precedent is not found, there is a search of the rules for the corresponding values of the chemical elements. The method of fuzzy logic is taken as a basis. The database structure is determined by the structure of the precedent and information used in the process of work.

The second block is a prediction of the optimum quantitative ratio of the chemical elements of CGI, based on the information inputted by a user without taking into consideration the influence of factors such as: the amount and the input order of modifiers, casting method, the methods of feeding of metal into the mold, kind of heat treatment, the rate and conditions of solidification of cast iron. We should consider not only the chemical composition of cast iron, but also the composition of inoculants.

The matter is that having the same chemical composition of the basic elements you can significantly affect the final result adding different kinds of modifiers. At the output from the block we'll obtain a quantitative content of the chemical elements to get the optimum properties of cast iron.

The fourth block is the definition of casting method, a method of feeding metal into a mold and heat treatment. The matrix of CGI directly depends on the casting method. Selection of the casting method depends on the following parameters: material, weight, complexity of construction and overall dimensions. Overall dimensions and casting weight often play a decisive role in choosing the method of manufacturing.

The fifth block is the definition of speed and conditions of solidification of cast iron. These criteria were chosen because the shape and quantity of graphite depend not only on the additives of modifiers, but also on the crystallization time. The nature of solidification of cast iron is determined by its composition and cooling rate. Slow cooling promotes graphite, accelerated cooling partially or completely suppresses graphite precipitation and contributes to the formation of cementite. The cooling rate depends on the wall thickness of the article; therefore, the mechanical properties of CGI depend on it either.

4. Conclusion

As the intelligent system consists of the stages (modules), it is possible to add or exclude certain modules according to the complexity of the problem; as a result, it expands the range of application of this intelligent system.

5. References

[1] V I Bolshakov Y I Dubrov, A N Tkachenko 2010 *The task of identifying the qualitative characteristics of materials on the basis of expert systems.* // Vestnik of Dnieper State Academy of Construction and Architecture : scientific and theoretical journal. - D.: PDACA - № 1. - P. 46-49

[2] V I Bolshakov Y I Dubrov, E J Zhevtilo 2009 *The research of efficiency and effectiveness of empirical empirical prediction of the quality characteristics of steel at the pre- stage of its creation.* / / Reports of the National Academy of Sciences of Ukraine: Scientific and theoretical journal. - N 9. - S. 103-107: Fig., Tab. - Ref. (8). - ISSN 1025-6415.

[3] V V Kurban *Improving the quality of the mechanical properties of hot-rolled sheet steel pipe using neural network modeling.* Materials and abstract of the author's dissertation (HAC) 05.02.23

[4] A D Chertov 2003 *Application of artificial intelligence systems in the metallurgical industry.*TsNIIchermet "Metallurg" . N 7

[5] I H Tukhvatulin, Y P Lankin *Forecast of properties of metallic alloys by means of neural networks*

[6] www.aiportal.ru/articles/neural-networks/

[7] Osovsky S 2004 *Neural networks for information processing*

[8] Kruglov V V Borisov V.V. 2002 *Artificial Neural Networks. Theory and Practice*

[9] Rutkovskaya D Pilinsky M., L. Rutkovski, 2006 *Neural networks, genetic algorithms and fuzzy systems*

[10]. Alekseeva L G *The influence of chemical composition of cast iron on its mechanical properties.* Section 15, "Instrumentation, control and diagnostics in technological processes, standardization"

[11] Bolshakov A E Lovushkin M Y 2011*Prediction of compositions of aluminum alloys by their properties* Mechanical engineering and life safety,№ 2, p. 37-43 ISSN 2222-5285

[12] F Sh Sharifyanov R F Mamleyev 2002 *Designing of castings*: Textbook / ; Ufimsk. gov. aviation. tehn. University -Ufa

[13] V T. Senchenko A A Yatsenko Y A Siniev 2001 *The production of castings from cast iron.* Part 1: Writing lectures. - SPb.: NWTU

[14] M O Matveeva B V Klimovich V V Klimovich E A Dvornikova 2013 *Simplex method of optimization of the cast iron with a given set of properties during the casting* / "System Technology" 2 (85) p.124 -131

[15] A A Radchenko A V Gal'chenko 2011 *Improving the quality of steel and cast iron castings.* Casting-2011: Materials of the VII International scientific and practical conference. p.154 -155

IOP Conf. Series: Materials Science and Engineering **69** (2014) 012024 doi:10.1088/1757-899X/69/1/012024

System analysis of the energy complex of engineering enterprise as a basic tool of effective energy management

N F Kashapov, I R Gil'manshin, I A Konahina

Kazan Federal University,
420008, Kremlevskaja Street, 18, Kazan, Russia

E-mail: chelny@kpfu.ru

Abstract. Approaches to system analysis of difficult energy objects of the machine building enterprise and its energy complex as a whole are considered. The role of system researches in the general structure of energy management of the enterprise is analyzed. Recommendations for choice of efficiency indicators and methods of revealing of technically recoverable losses of energy resources are offered.

1. Introduction

Improving energy facilities of engineering enterprise that combine dozens of pieces of equipment is a complex task that is possible to solve only on the basis of system analysis methodology [1, 2]. The complexity of the task is defined with hierarchical rank the selected object of study: apparatus □ setting □ technological system, etc. The higher the level of the object is, the wider the range of possible solutions for its improvement.

2. The algorithm of the research

The algorithm of the system of research involves a number of sequential subtasks:

1. Analysis of the effectiveness of the original object vigorously to the selected criterion or set of criteria.
2. Structural analysis of the object with the introduction of elements hierarchy of the equipment included in its composition. The hierarchy is built on the degree of importance of a particular element to obtain the final result. The dominant and dependent components of the system are pointed out (including the dependent chain elements).
3. Decomposition of a complex object.
4. Definition of a set of measures aimed at improving the facility.
5. Introduction of new pieces of equipment in the structure of the object.
6. Mathematical modeling of the object.
7. Synthesis of a new object
8. Optimization of synthesized object.

The international standard of ISO 50001:2011 "Energy Management Systems" [3] identifies a number of functions that must be performed at the plant in order to ensure effective management of the expenditure of energy .:

- The development of methodology and updating of energy analysis;
- Construction of the energy balance of the enterprise on the basis of measurements and other data;
- Determination of areas (equipment, subsystems, systems) meaningful use of energy;
- Identifying and ranking of opportunities for improving the level of enterprise energy efficiency;
- Planning the implementation of energy saving measures;
- Compensation of plans if the current results of energy analysis show a need for it;
- Search of mechanisms and means of implementation activities, including energy service contracts.

From the comparison of the features of the system analysis and energy management functions can be concluded that the system analysis is a tool, which is based on the possible development of a methodology energy analysis of large production units, including engineering enterprises.

An important issue in enterprise energy management system is the selection of energy efficiency indicators. As shown in [3, 4] indicators are selected for different enterprises in different ways, based on the guidance of the goals of the enterprise. It can be simple parameters, such as the amount of energy consumed per year or the number of energy consumed per unit of product. But also more complex parameters and the models can be applied.

3.Conclusion. In practice, it is usually limited to simple performance indicators, which are not always informative. For example, the performance indicators defined on the basis of the energy balance of the energy facility, reflect the amount received and the useful energy consumption. The issue of quality of energy used in this way is not affected. Meanwhile, the potential of energy resources (e.g. steam pressure) in this site may be used not enough. To determine the efficiency of energy resources and the extent of its use in the process of energy is rational to apply the parameters of thermodynamic efficiency of the object [2].

Methods for determination of compliance with quality energy resources, for example, the method of compliance with the temperature [5], allows to detect technically avoidable loss of efficiency of energy resources, including direct loss of energy, and also to identify the processes, implementation of which will eliminate these losses.

4.References

[1] Melent'ev L A 1983 *System studies in engineering. Elements of the theory, development trends.* 2nd Edition - M: Nauka

[2] Nazmeev J G Konahina I A 2001 *Organization of energy technology complexes in the petrochemical industry.* – M.: Publishing house of MEI

[3] *The international standard ISO 50001: 2011. Energy Management Systems* - Requirements with guidance for use

[4] *Practical tips for energy saving in the rubber processing industry.* – Good Practice Guide No 262, Energy Efficiency Best Practice Program, ETSU, Great Britain – Crown copyright 1999

[5] V G Grigorov, V K Nejman, S D Churakov and others 1987 *Disposal of low-grade thermal waste energy in chemical plants.* / – M.: Chemistry

Optimization of industrial steam supply and steam-and-condensate farming of machine building enterprise

I A Konahina[1], N F Kashapov[1], I R Gil'manshin[1], R R Ganiev[2]
[1]Kazan Federal University,
420008, Kremlevskaja Street, 18, Kazan, Russia
[2]Kazan State Power Engineering University,
420066, Krasnosel'skaja Street, 56, Kazan, Russia

E-mail: chelny@kpfu.ru

Abstract. The article studies efficient control methods of steam condensing economy of the machine building enterprise. There are recommendations about development of complex decisions based on indicators of energy, technical and economic efficiency.

1. Introduction

Currently, in the energy of Russian Federation a key strategic direction is energy saving at all levels. Particularly acutely it is concerned energy intensive industries. Machine building enterprises are the largest consumers of steam. Basically it is consumed for technological needs, but also providing of "domestic" consumers with steam is not an exception – on heating, ventilation and hot water (DHW).

The steam for enterprise is released from the industrial CHP, as a rule, with the help of one steampipe where the maximum pressure of the system is maintained, and the real consumption of steam by numerous consumers is realized at lower settings. Application of reduction plants or reducing-cooler settings (PRDS) in these cases leads to a loss of exergy of energy source that makes us to look for solutions for eliminating such losses [1,2].

The use of steam for heating needs leads to formation of high pressured condensate, towards of atmospheric, which also represents the reserve of energy saving. Except condensate of thermal needs the exhaust steam with the forging hammers' pressure of 0.2-0.3 MPa, presses and other installations with steam drive can be used.

2. The basic part

Based on the analysis of the structure of machine building enterprises' power balance the following ways of improving efficiency in the use of steam and condensate were identified:

1) Installation of low pressure turbines PRDS (Image 1);

2) Applying of recompression to recover the capacity of waste steam;

3) Accumulation of steam and heat energy;

4) Organization of effective systems of collection and return of steam condensate, as well as re-use of energy supply (regeneration, recompression of flashing steam and so on.);

5) Optimization of steam supply systems, including steam condensate farming enterprises

Image 1 shows a diagram of the replacement of pressure reductionin PDRS on its expansion process in the turbine. Using of this scheme allows generating of 98.3 kWh of electricity per 1 tone of steam for the specified mode parameters and configuration.

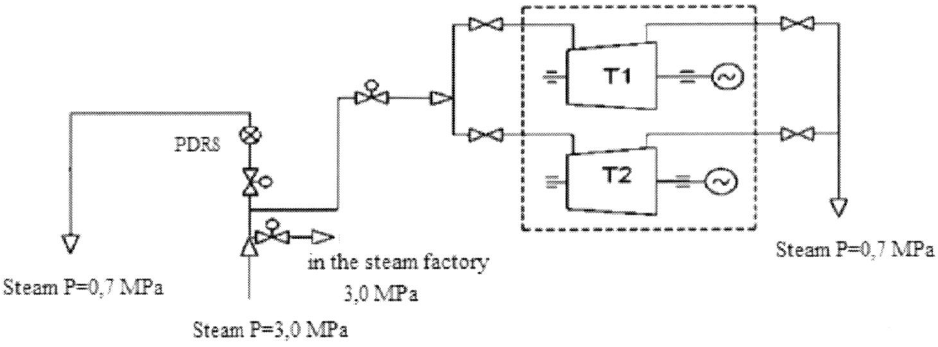

Fig.1. Cogeneration bundle
T1, T2 - turbines P-1,6-2,8/0,7

After using the steam in power technological processes low potential renewable energy resource (RER) of waste steam is formed, the continued use of which is difficult because of its low pressure and temperature, so such a heat-transfer fluid is usually released into the atmosphere – either directly or via the circulating water system. To increase the pressure of secondary steam to an acceptable level it is advisable to use recompression methods based on different types of compressors, including steam jet. During this process the additional costs of energy resources – electricity or steam of industrial settings are produced.

Disposing of waste steam, steam jet compressor (2a) at the same time works as PDRS, reducing high-grade steam parameters to the calculated working values. The steam with the operating parameters is then delivered to the consumer.

Gas generator installation (im. 2b) compresses the steam to the designed pressure. The thermal using of pressurized steam effect of fuel economy reaches 3-6 times value compared to the produced electricity costs on the compressor's drive.

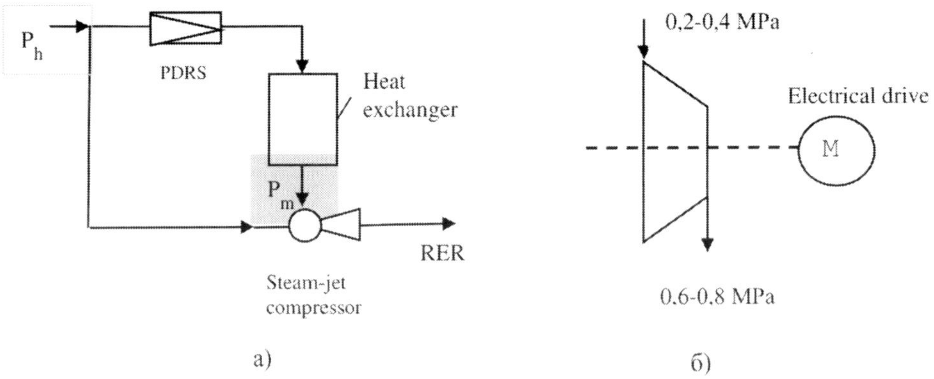

Fig. 2. Steam recompression by
a) Steam-jet compressor, б) turbo compressor

IOP Conf. Series: Materials Science and Engineering **69** (2014) 012025 doi:10.1088/1757-899X/69/1/012025

Daily schedules of steam consumption at the enterprise are characterized by unsteadiness. Accumulation of the steam (Fig. 3) lets to stabilize the schedule and ensure the effectiveness of steam supply systems and steam consumption teamwork [2]. Charging the battery takes place in a period of excess income of high pressure steam and discharge in a period of inadequate intake of medium pressure steam.

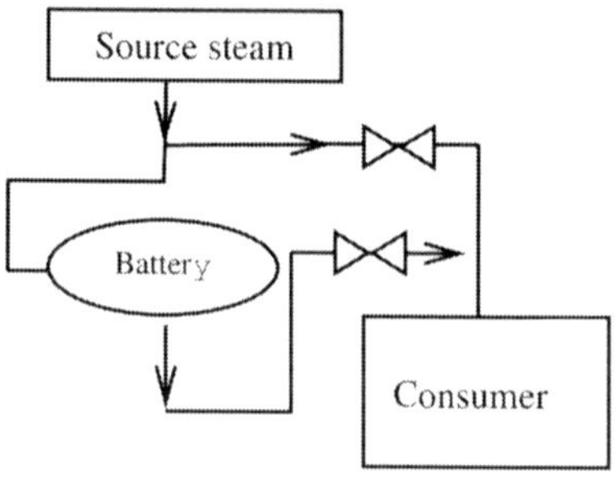

Fig. 3 Enabling the battery in steam consumption system of the enterprise

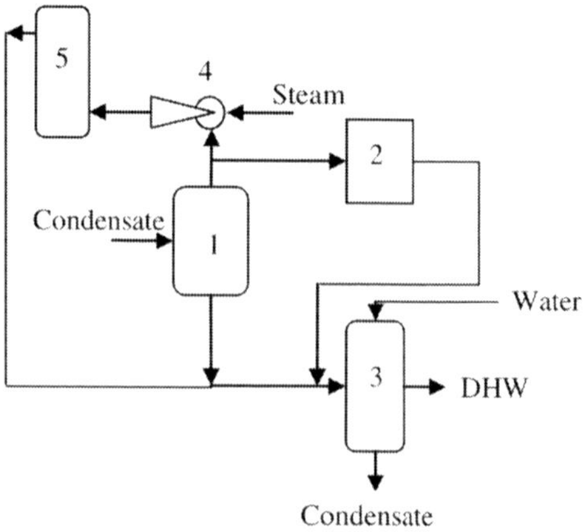

Fig. 4 Heat recovery condensate of different pressure

For the purpose of beneficial use of RER steam condensate a complex of measures is proposed (Fig. 4), which includes the installation: Expansion tank for emission of flash steam; – Steam jet compressor to pressurize the recycling steam to the desired pressure; – Waste-heat exchangers -utilizers for use RER of residual condensate.

Condensate after using enters the expansion tank 1, generated flash steam is fed into the process unit 2, and a steam jet compressor 4 (to increase the capacity of the steam). Steam after steam jet compressor enters the technological unit – the consumer 5. Condensate from 1, 2, 5, enters the heat exchanger 3, where also gives its residual heat to system of domestic hot water (DHW).

The task of improving the steam-and-condensate farming of machine building enterprises is complex and has many solutions. Search of the most profitable of these methods requires the use of system analysis, structural modeling and optimization [2, 3]. The challenge of energy-economic optimization in this case is to find the set of technical solutions and operational parameters that correspond to the maximum achievable fuel economy B_e at imposing restrictions caused by the peculiarities of the applied technologies and equipment [2].

$$B_e = \sum_{i=1}^{i_k} (\Delta B_D + \Delta B_N)_i \; ; \qquad (1)$$

where – ΔB_D is change of fuel equivalent of steam consumption by enterprise from external sources t/h; ΔB_N is change of fuel equivalent cost of electricity from external sources; i – the serial number of the event; i_k – the total number of events.

However, as shown in [2], optimization of complex according to the energy parameters represents idealized problem, in which does not take into account the level of capital and operating costs of a particular event. Meanwhile, the decision to implement complex in practice is adopted on the basis of technical and economic parameters of the project. Therefore, the problem (1) must be solved together with the task of finding the maximum economic benefit ΔE_c or minimum payback period τ_r (2) of the energy-saving measures, taking into account the cost of providing the required conditions of reliability and safety upgradeable complex:

$$\tau_r = \frac{\sum_{i=1}^{i_k} (\Delta S_c)_i}{\sum_{i=1}^{i_k} (\Delta E_c)_i} \quad ; \tag{2}$$

where ΔS_c - is capital costs on event, rub.

3. References

[1]. Levin M S *Use of steam and secondary steam and condensate* 1971 M.: Energy

[2]. Nazmeev Ju G, Konahina I A *Thermal energy systems and energy balances of the industrial enterprises* 2002 M.: Publishing house of MEI

[3]. Bass M S , Batuhtin A G *Integrated approach to optimize the functioning of modern heating systems. Thermal Engineering*, 2011 №8

The method of manufacture of nylon dental partially removable prosthesis using additive technologies

R N Kashapov[1,2], A.I. Korobkina[1] , E.V. Platonov[1] and G.T. Saleeva[3]
[1] Engineering Institute, Kazan Federal University, Kazan, 420008, Russia
[2] Laboratory of Radiation Physics, Kazan Physical-Technical Institute of the Kazan Scientific Center of the Russian Academy of Sciences, Kazan, 420029, Russia
[3] Kazan State Medical University, Department of Orthopedic Dentistry, Kazan,420012,Russia

E-mail: kashramiln@gmail.com

Abstract. The article is devoted to the topic of creating new methods of dental prosthesis. The aim of this work is to investigate the possibility of using additive technology to create nylon prosthesis. As a result of experimental studies, was made a sample of nylon partially removable prosthesis using 3D printing has allowed to simplify, accelerate and reduce the coat of manufacturing high-precision nylon dentures.

1. Introduction

Removable prosthesis is one of the most popular areas of dentistry. This is explained by the statistics of clinical cases. For example, 30% of the total population of the UK are partially removable dentures. In the U.S the expected total demand for fixed and removable prostheses by 2020 pledged as 115% of the current means-tasted [1].One of the materials are finding increasing application is nylon. Thia is due to the following factors: it does not cause allergic reactions, has a stable dye, and therefore retains its appearance even after prolonged use. Has less weight and high flexibility compared with other materials, which reduces the period of adaptation to the prothesis.

The most popular are partially removable dentures due to the lack of necessary grinding the adjacent healthy teeth and their use as a temporary structure before fixed prosthesis [2].

Modern technology create nylon dentures includes a large number of complex preparatory stages. Namely, hand modeling technician on the plaster model of the prosthesis wax, preparation of the cell for extrusion of nylon and super plasters for sagipsul system removable prosthesis(over the path sprue),using a thermo-press for extrusion of nylon when heated to 250°C,the use of a special cartridge, which is placed in a heat press for 11 minutes, followed by the procedure of pressing the shape of the prosthesis and cool down the cell, processing by means of cutters, polishers, polishing filename, brushes, grinding wheel. All of this shows the complexity and duration of the manufacture of nylon prosthesis and pushes on the need to develop new methods of production [3]. In recent years, an increasing distribution are additive technology of formation of products. Therefore, the aim of this work was study the possibility of using 3D printing FDM method to create nylon prosthesis.

IOP Conf. Series: Materials Science and Engineering **69** (2014) 012026 doi:10.1088/1757-899X/69/1/012026

2. Experimental

Modern method of fabricating a partially-removable dental prostheses are not able to give the technique a complete idea about the individual peculiarities of the structure of the dental apparatus of the patient.

Kazan Federal University in the Centre for precision technologies dental prosthesis developed a technology for manufacturing dental partially removable dentures made of nylon with 3D technology. The use of additive technologies can make denture most anatomically correct, thus avoiding discomfort and malocclusion of the patient.

At the initial stage of manufacturing partially denture is removed and the impression denture the patient's jaw, then in the laboratory made a plaster model. Received a plaster model scan using dental scanner Activity 875.This scanner is designed for holistic scan of the jaws using craniofacial coordinates. Model pre-configured in the articulator (with help of masticator), can be located and oriented in the scanner through the use of appropriate magnetic intercalation plates. Increased scan area and a full 2-axis automated scanning system that provides highly accurate data model of the jaw.

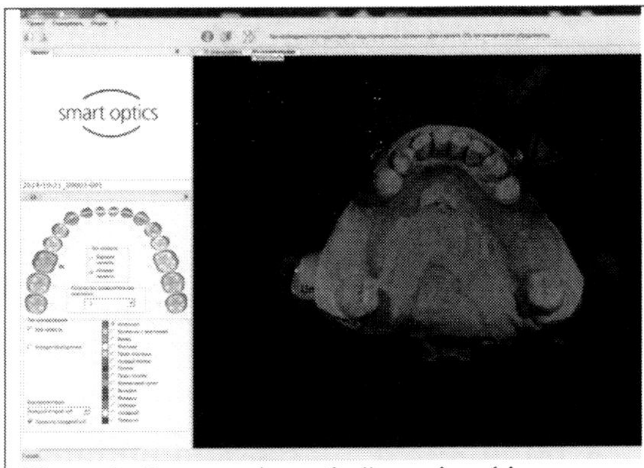

Figure 1- Pre scanning a 2-dimensional image

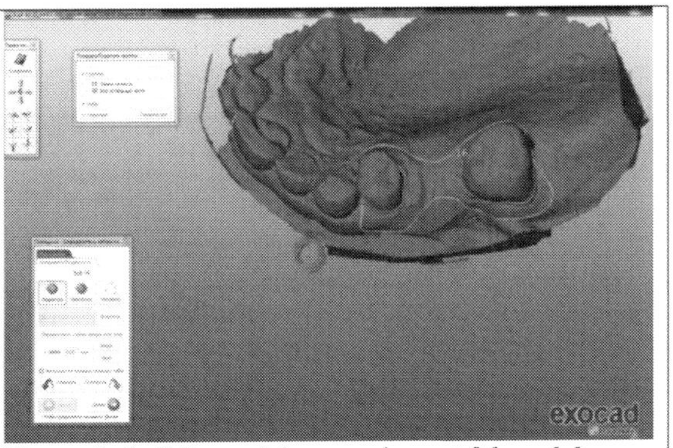

Figure 2- building of frame splace at 3d model

Pre scanning a 2-dimensional image of the fixed model. This image serves as the basis for determining the position of the scan. The scanned 2-dimensional picture is processed using the scanner software Activity 875, the scanning procedure is generated 3D model. The obtained model defined boundary position of the frame partially removable prosthesis (figure 2).On selection of the simulated framework of the future prosthesis, formed clamp and saddle. Determining the boundary of the adjoining clasps, the program automatically fills in the border of the prosthesis shown in figure 3.The program Exocad have a large number of tools to ensure the desired geometric shape of the prosthesis. Thanks to the digital modeling clasps made a nylon maximum snugly in contact with the gums, just repeating its shape and providing a reliable fixation in the oral cavity.

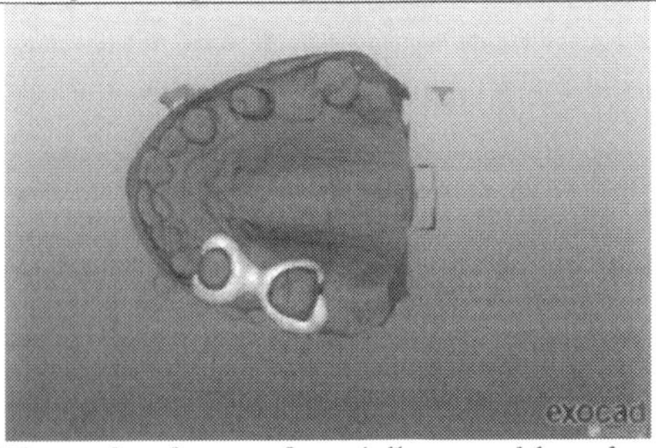

Figure 3 – frame of partially movable nylon prosthesis

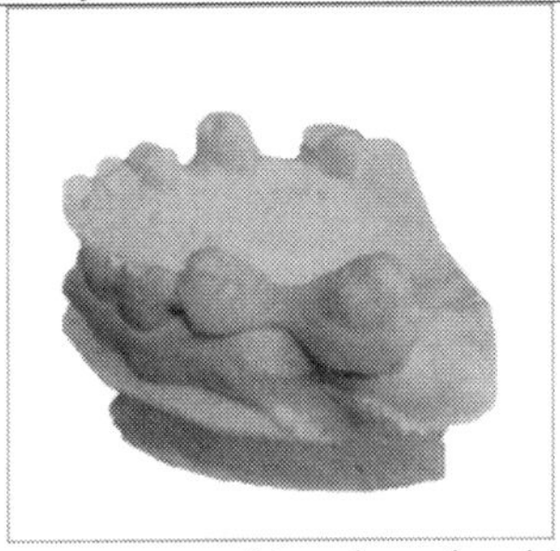

Figure 4 – prototype of bends and saddle-backed part of prosthesis.

After the process is complete, the simulation results are stored on your computer's hard drive in STL format. Fabrication of the prosthesis is carried out using 3D printer Picaso Builder. This printer works on technology of extrusion plastic-i.e, the model is obtained by layer-by layer deposition of very thin layers of molten nylon. The finished product thus obtained is accurate, practical and durable. The printer uses the Polygon program, which loads the fike into STL format with the model of the prosthesis. Print time dental of partially removable prosthesis is about 13 minutes. The frame of the prosthesis reprinted from material Nylon Co-Polimer 645 (made in USA). Value 1,500 cm^3 of material is a 3,000 rubles, for the manufacture of the frame by one unit requires 2 cm^3.The cost of this wireframe is 4 ruble, which is much cheaper modern methods of manufacture frames and requires further research.

After printing nylon clamp should be mechanically remove material support. Next,to give a smooth aesthetic forms were carried out different finishing methods: inkjet surface treatment was performed on the installation of APO-5U (Aveyron, Ekaterinburg),the processing in pairs of acetylene. The prototype clasps and saddle-shaped part of the prosthesis is presented in figure 4.

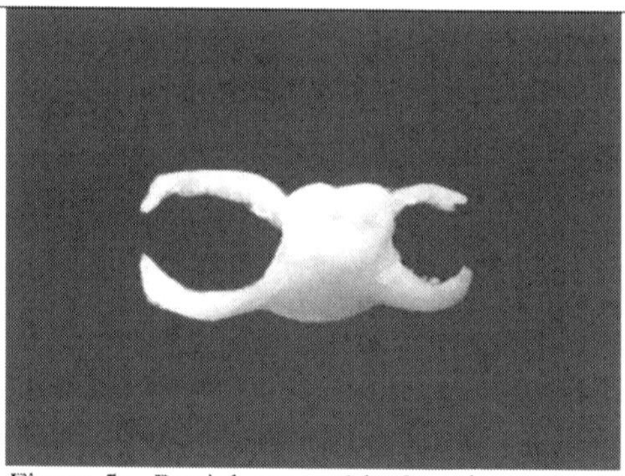

Figure 5a- Partial removable dental prosthesis made of nylon, printed on a 3D printer Picaso Builder.

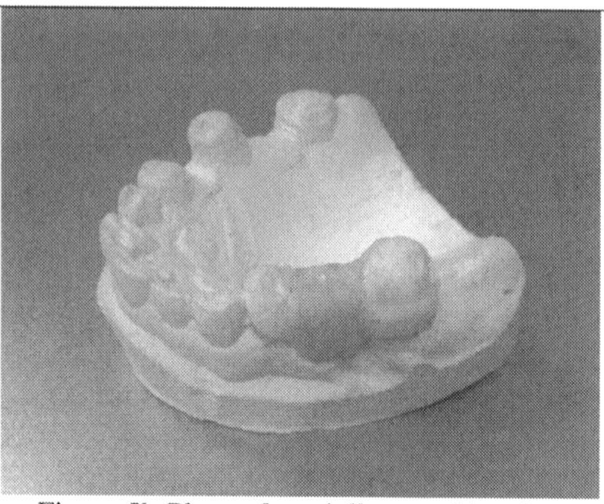

Figure 5b-Place of partially movable ntlon prosthesis on the jaw

Figure 5 a, b - Partially - removable dental prosthesis made of nylon, printed on a 3D printer Picaso Builder

Printed nylon frame is further transferred to the dental technician. Installation of the dentition on the saddle is carried out by hot polymerization (figure 5 a, b). Developed manufacturing technology clasps and saddle part no adverse effects on the prosthesis bed. This method of implants using additive technology to create the most accurate prosthesis for a short period of time.

The development and use of additive technologies in dentistry greatly simplify the process of creating dentures. Numerical modeling allows to take into all the structure of the oral cavity.The technology of fabrication of a prosthesis made of nylon allows to create nylon dentures quickly, at low cost, with no large outlay of recources. The accuracy of the prosthesis is increased through the use of digital technology and 3D prototyping.

Given all modern disadvantages of prothesis, this method is an alternative proposal to allow time-consuming milti-step process to simplify and improve quality of partially-removable nylon denture.

Acknowledgments

This work was financed by grants of Kazan (Volga region)Federal University of the state of tasks in the sphere of scientific activity.

References

[1] Nicholas J.A. Jepson Removable Partial Dentures 2006
[2] Tregubov I.D. Application of thermoplastic materials in dentistry/I.D. Tregubov Liter,Kirilenko.-M,2007-s.
[3] Plasma application of protective polymer-powder coatings to ultrasonic sensors.BiomedicalEngineering.2011.T45(5) 138-200

Providing normal conditions of lubricating of diesel engine during its operation

A T Kulakov, A A Gafiyatullin, E P Barylnikova
Kazan Federal University,
423800, 68/19 Mira avenue, Naberezhnye Chelny, Russia

E-mail: chelny@kpfu.ru

Abstract. This paper analyzes the experience of operating an internal combustion engine crankshafts and bearings. The influences on failure depressurization in the lubrication system are evaluated. Means of maintaining pressure in operation are proposed and estimated.

1. Introduction
When operating engines of KAMAZ there is a change that occurs lubrication conditions of crankshaft bearings, significantly reducing the reliability and limiting technical resource [1]. One of the major factors causing engine failures is to reduce the oil pressure in the lubrication system [2, 3, 4].

2. Basic part
Statistical data on engines KAMAZ-740 show (Table 1, Figure 1) that during the operation in the lubrication system there is a decrease of pressure on operating time, which can be represented as an exponential probability model [5]:

$$P = P_0 \cdot e^{-bl}, \qquad (1)$$

where P_0- pressure in the lubrication system at the beginning of operation; b - coefficient characterizing the slope of the curve reducing the pressure in the lubrication system; l - working hours

Table 1 - Dependence of pressure in the lubrication system of the engine KAMAZ-740 on the mileage at nominal and minimum frequency.

Frequency, min-1	Operating the vehicle mileage, thousand km						
	20	40	60	80	100	120	140
2600	0,39	0,34	0,35	0,29	0,24	0,23	0,23
600	0,21	0,17	0,16	0,12	0,07	0,09	0,07

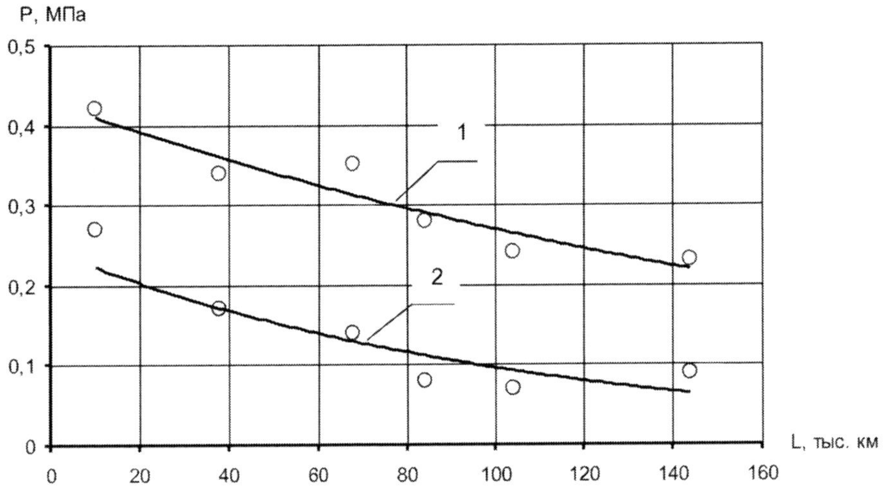

Figure 1 - Dependence of pressure in the lubrication system of the engine KAMAZ-740 on operational vehicle mileage: 1 - at n = 2600 min-1, 2 - for n = 600 min-1.

Effectiveness of the lubrication system depends largely on the oil pump performance and reliability of its work [6]. The required pump capacity is determined by the amount of oil needed for distribution to consumers. The reason for reducing the pressure in the engine lubrication system at nominal mode to P = 0,2 ÷ 0,3 MPa at 100 ÷ 140 thousand km may be the lack of performance of the oil pump. In normal operating conditions may occur in critical modes of the connecting rod bearing lubrication, caused by reduced oil pressure in the lubrication system [7, 8].

The productive capacity of the oil pump engine KAMAZ-740 is 85 l /min in the main section, and 8 l / min in the radiator section at nominal conditions [9]. The sufficiency of such magnitude performance was determined in the study of connecting rod bearing lubrication conditions under test bench manufacturer [10, 11].

For this pressure measurement was carried out in the lubrication system at different temperatures and oil engine operating modes (no load, with load) with locked valves (differential and the safety did not work for oil drain) [12].

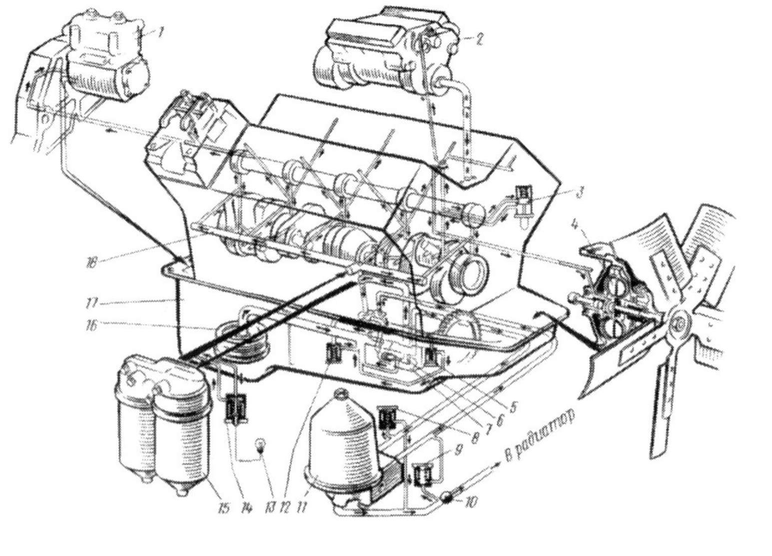

IOP Conf. Series: Materials Science and Engineering **69** (2014) 012027 doi:10.1088/1757-899X/69/1/012027

1- compressor, 2 - high pressure fuel pump 3 - switch of fluid coupling, 4 - sifter, 5, 12 - safety valves, 6 - valve of lubrication system 7 - oil pump, 8, 9 - centrifugal filter valves, 10 - valve of switching of oil radiator, 11 - centrifugal filter, 13 - alarm lamp of clogging of filter 14 - full-flow filter bypass valve, 15 - the full-flow oil purification filter, 16 - oil receiver, 17 - Carter, 18 - the main oil gallery

Figure 2 - The engine's oiling system KAMAZ-740

Figure 3 shows the obtained maximum performance pressure in the lubrication system of the engine depending on the temperature without load. It is seen that by increasing the temperature, the pressure in the lubrication system is reduced significantly, which indicates an insufficient capacity of the oil pump. When the oil temperature is 90 ÷ 100 ° C appears performance margin (at P = 0,45 MPa) at a frequency of n> 1700 min-1 with no load. Oil at temperatures over 100 ° C margin performance is virtually absent (Figure 3).

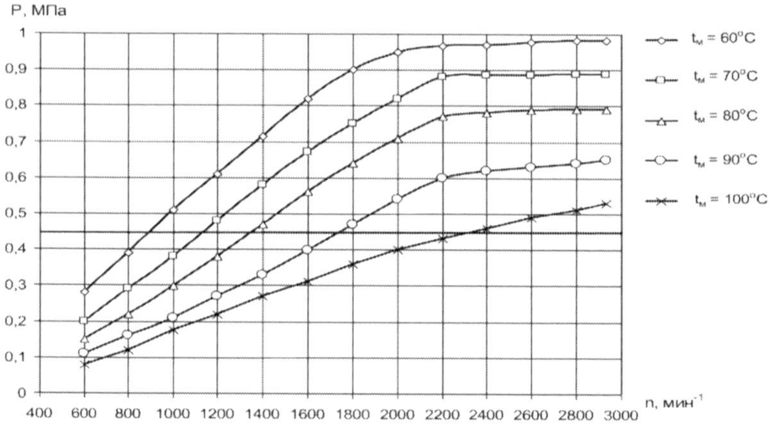

Figure 3 - Characterization of pressure in the lubrication system according to the no-load temperature differential locked and the safety valves.

With the engine running at 90 ° C margin performance at idle appears when the engine speed is more than 1,650 m −1, and at full load at 2250 min −1 and is less than 15 L / min at a frequency of 2930 min −1.

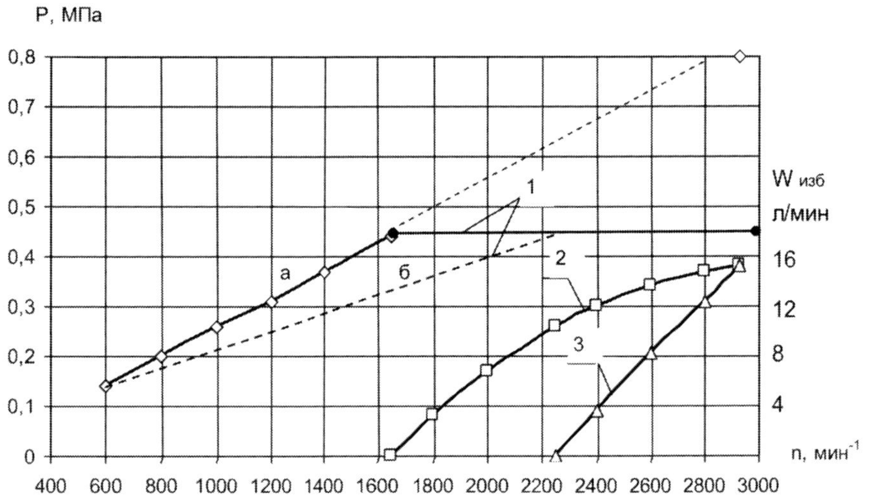

1 - Reproducible characteristic of pressure with operating (unlocked) valves (a - at Tm = 90 ° C without load, b - at Tm = 90 ° C with a load), 2 - stock performance at idle 3 - stock performance with load

Figure 4 - Determination of reserve oil flow

When acceptance tests in factory conditions there are engines with pressure below P = 0.1 MPa at the speed of 600 min-1 and below P = 0.45 MPa at 2600 min-1. Engines with a flatter characteristic 6, 7 (Figure 5) are removed from the test, as they have no stock performance. However, sometimes there is release of the motor without performance reserves.

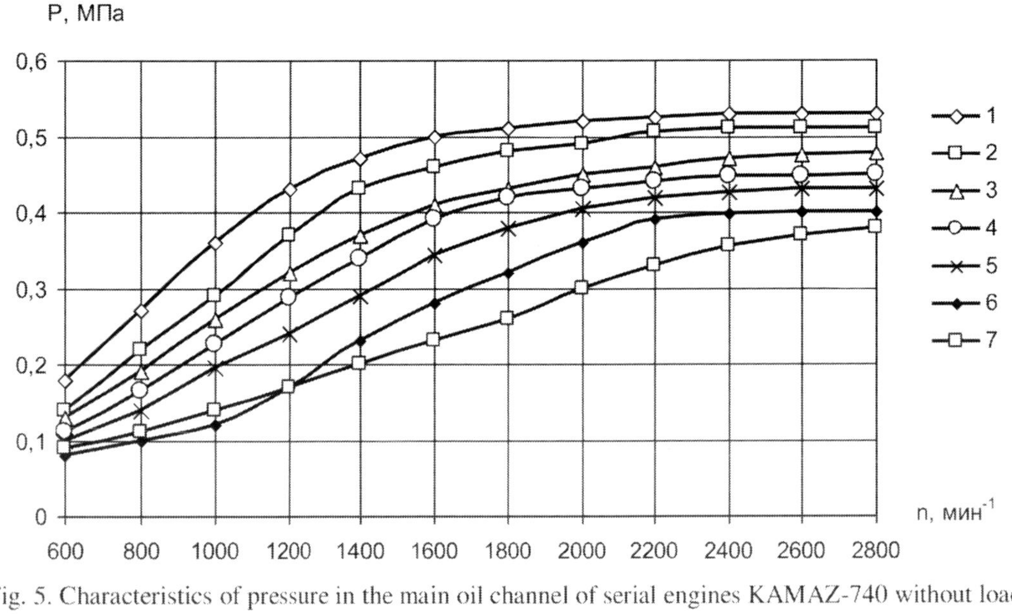

Fig. 5. Characteristics of pressure in the main oil channel of serial engines KAMAZ-740 without load (Tm = 80 ... 90 ° C)

Operating the engine without the stock performance of the oil pump reduces the pressure in the lubrication system in operation long before the establishment of a resource [13].

Carry out the calculation of the resource pressure in the engine lubrication system.

IOP Conf. Series: Materials Science and Engineering **69** (2014) 012027 doi:10.1088/1757-899X/69/1/012027

Reducing the pressure in the lubrication system of the engine during operation due to the increase in the bearing gaps and occurs according to an exponential (1).

According to data obtained from the operation (Figure 1), we define the coefficient "b".

Initial data: l = 140 thousand km P = 0.23 MPa; P0 = 0.45 MPa.

We find that b = 0,0049.

Critical lubrication regime of connecting rod bearing at n = 2600 min-1 without load is Pct. = 0.29 MPa (rupture of the oil flow). Given this resource engine to achieve the critical pressure of the formula (1) we obtain for the initial characterization with headroom (see Figure 4)

$$l = \frac{1}{b} \cdot \ln \frac{P_0}{P_{кр}}, \qquad (2)$$

where P0 = 0,75 MPa; Pkr = 0.29 MPa, and therefore following is calculated:

$$l = \frac{1}{0{,}0049} \cdot \ln \frac{7{,}5}{2{,}9} = 193 \text{ thousand km.}$$

If the engine comes in operation without reserve capacity (P0 = 0.45 MPa at rated power) its resource is reduced and the life of the engine is:

$$l = \frac{1}{0{,}0049} \cdot \ln \frac{4{,}5}{2{,}9} = 87 \text{ thousand km.}$$

Determine the need pressure at nominal conditions to achieve resource engine 300 thousand km. From (1) we get:

$$P_0 = P_{кр} \cdot e^{bl} = 2{,}9 \cdot e^{0{,}0049300} = 1{,}3 \text{ МПа}$$

This pressure corresponds to the characteristic shown in Figure 6.

Setting pressure reducing valve should be raised to 0,6 ÷ 0,7 MPa in order to shift the characteristics on the critical pressures.

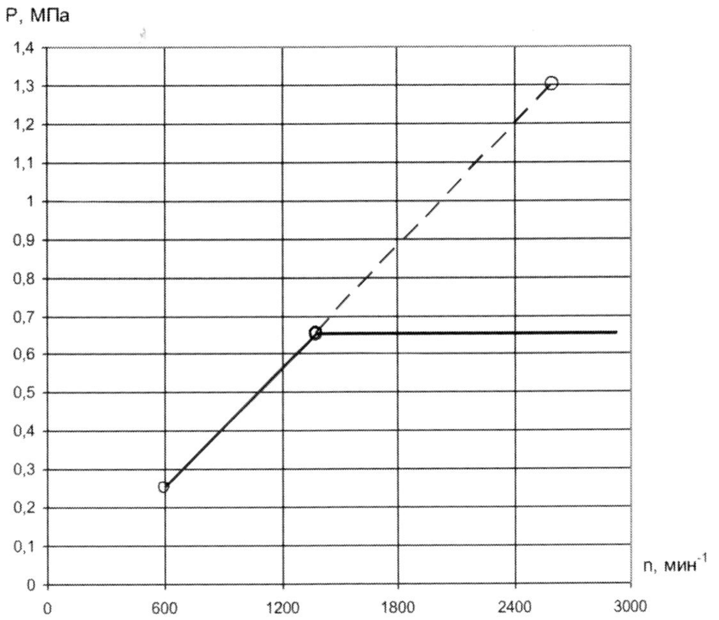

Fig. 6 Required characteristic pressure in the lubrication system of the engine to achieve resource 300 thousand km

In modern engines, starting with the KAMAZ-740-11-240 lubrication system has been changed. The oil pump with enhanced performance of up to 150 l / min (Fig. 7) is used [14].

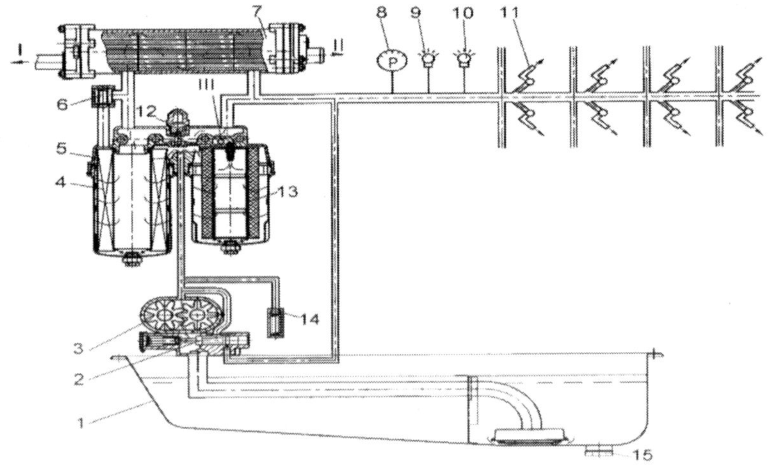

1 - oil casing; 2 - valve of lubrication system; 3 - the oil pump; 4 - full-flow filter element; 5 - The oil filter; 6 - relief valve; 7 - oil-water heat exchanger; 8 - oil pressure gauge; 9 – indicator of oil pressure; 10 - oil temperature alarm; 11 - piston cooling nozzles; 12 - thermostatic valve; 13 - part-way filter element; 14 - safety valve; 15 - oil drain plug; I - coolant return ; II - a cooling fluid inlet; III- plums in a oil casing.

Figure 7 - The engine lubricating oil system KAMAZ-Euro 1

Using high-efficiency pump allowed slowing down the rate of reduction in operating pressure (Fig. 8)

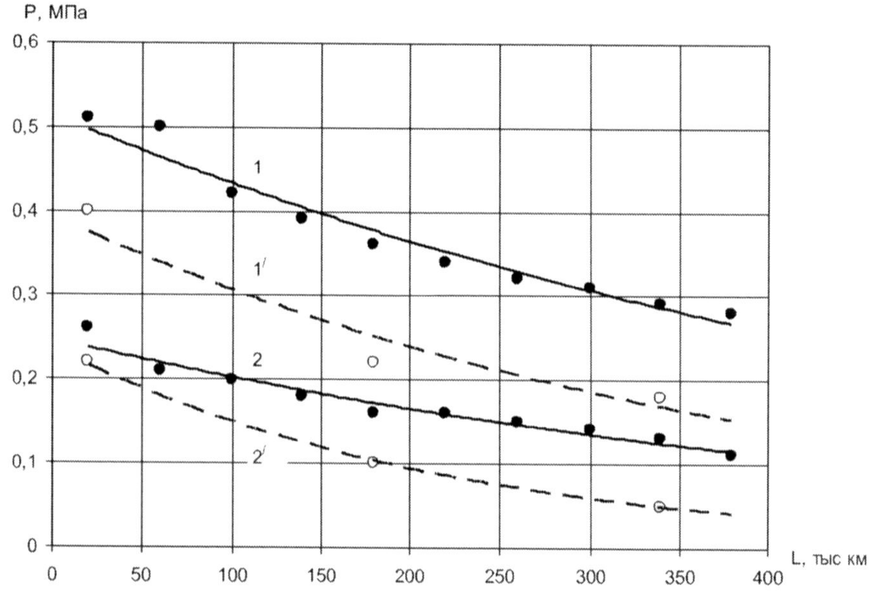

Figure 8 - Dependence of pressure in the lubrication system of the engine KamAZ use: unprimed - KamAZ-Euro; primed - KamAZ-740 (1 - for n = 2200 min-1, 2 - for n = 600 min-1)

3. References

[1] Smirnov V G 1980 *Increased durability of parts of car engines by improving the design of lubrication systems* M. NIINavtoprom,. 59 p

[2] Denisov A S 1981 *Analysis of the causes of service failures of connecting rod bearings of engines KAMAZ-740* AT Fists / / Engine building Number 9.pp 37-40

[3] Denisov A S 2005 *Analytical study of changes in the conditions of rod bearing lubrication during operation* / Saratov: Sarat. Vesnik. Reg. Techn. Univ.. Number 3. pp 69-75

[4] R T and Kashapov N F 2001 Iz. VUZ. Aviatsionnaya Tekhnika **2** 80

[5] Denisov A S 1999 *Foundations for operating and repair cycle of cars* Saratov: Sarat. Reg. techn. Univ 352 p

[6] Grigorev M A 1978 *Ensuring the reliability of the engines* M.: Standards Press 324 pp

[7] Barylnikova E P 2011 *Investigation of the conditions of the bearings of the crankshaft of the engine* Proc. scientific. tr. "BULLETIN OSU» № 10/oktyabr. - Orenburg: SEI OSU pp 135-138

[8] Barylnikova E P 2013 *Kinetics of the lubrication of diesel, dissociation and formation of oil flow imbalance/* Advanced technologies in transport systems / Sat Art. Eleventh Intern. scientific-practical . conf. - Orenburg Reg. Univ, pp 49 -54

[9] Denisov A S, A T Kulakov 2007 *Ensuring the reliability of automotive engines* . Saratov: Sarat . Reg. tehn. University, 350 p

[10] Perelsky A K 1993 *Heavy trucks KAMAZ* M.: Higher school, 303 p

[11] Barylnikova E P 2012 *Apparatus for studying the conditions of lubrication of the crankshaft /* Tractors and farm machinery № 12 - pp 38-40

[12] Barylnikova, E P 2012 *Experimental setup for the study of lubrication to the bearings of the crankshaft of the engine* Transport and transport- technological systems : proceedings of the International Scientific and Technical Conference . - Tyumen TSOGU pp 118 -122
[13] Barylnikova EP 2013 Influence of operational factors on the lubrication conditions of crankshaft bearings of automotive engines Tractors and farm machinery - № 5 pp 33-36
[14] Azamatov R A 2007 *Restoring parts of the power unit KAMAZ- 740.11 -240 (EURO- 1)* JSC Kamaztechobsluzhivaniye Naberezhnye Chelny 307 p

Innovations of modern system of vocational guidance in technical high school

T B Kurbatskaya, T I Bychkova
Kazan Federal University
423800, Republic of Tatarstan, Naberezhnye Chelny, 68/19 Mira avenue

E-mail: chelny@kpfu.ru

Abstract. This article is devoted to the description of system of modern innovative forms of vocational guidance in a higher educational institution of a technical profile.

1. Introduction

Technological modernization of Russia is impracticable without development and improvement of engineering education which has to be based on the best traditions of the Russian engineering school. In the 90th years of the XX century as many researchers state, the Russian engineering school in many respects lost the former positions. The rupture of communication of science, education and the industry led to departure of the most talented engineers in other spheres of action, many groups of researchers and developers broke up. Participation of technical and technological colleges in development of the industry, space and defensive complexes was decreased considerably. Nowadays preparation of technical shots and their employment not fully conform to modern requirements. According to the results of commission session at the President of the Russian Federation on modernization and technological development of economy of Russia instructions are approved the by D. A. Medvedev which realization is urged to promote support and development of the Russian engineering education. In them, in particular, the important part is assigned to creation of effective system of vocational guidance of youth; to increase of prestige of engineering professions; to professional development of engineering shots and quality of their preparation; to strengthening of state-private partnership in questions of training of specialists; to introduction in practice of continuous monitoring of requirement for technical shots on economy branches taking into account inquiries of a labor market; to social and economic support of teachers of educational institutions with a technical profile, the students who are training on engineering specialities; to development of material base..

2. Basic part. According to the Federal target development program of education for 2011-2015 [1] at the Institute Naberezhnye Chelny is developed the system of vocational guidance and prevocational training in educational institutions for increase of their motivation to the subsequent work. In it are provided measures for ensuring increase of appeal of training for educational programs of higher education with an engineering profile; increases of level of technological education of pupils, restoration of necessary volumes of technological training of pupils in all forms of comprehensive school; profound pre-university training of pupils in subjects of a natural-and-technological science by development of profile classes. Our purpose was creation the system of the educational space helping the personality at different levels of training to develop inclinations of technical and design abilities for further professional activity at the automobile branch [2]. In this system we sought for realization:

- concepts "Education during all life";
- systems of the formation of a professional orientation;
- individualization of a choice of an educational trajectory;

IOP Conf. Series: Materials Science and Engineering **69** (2014) 012028 doi:10.1088/1757-899X/69/1/012028

- orientations to a concrete type of the follow-up professional activity according to requirements of employers;
- uses of the innovative methods, pedagogical technologies and training techniques;
- possibilities of the optimization of the system and processes in its structure on the basis of the analysis of carried-out monitoring;
- systems of an assessment of the quality of the work.

The system of the professional orientation work with pupils of schools and students is following at present :

1 . Professional orientation diagnostics:
- Diagnostics of professional intentions of pupils of 9 and 11 forms (it is carried out together with the Center of testing and developments (Moscow) since 2007 by means of the computer Proforiyentator program. This testing helps pupils to reveal professional tendencies, to choose a specialized secondary or higher educational institution, training specialization in higher education institution, to define a suitable profile of training in the senior classes of school (in November, 2012 this diagnostics passed 189 pupils of profile schools) [3];
- Diagnostics of professional and important qualities of the personality by means of the device of system diagnostics "Aktivatsiometr ATs-9". During the period since September, 2012 research on the device of system diagnostics there passed 168 people [4].

2 . Professional education:
- the organization of excursions in the museum of history of higher education institution for pupils of general educational institutions in Naberezhnye Chelny;
- the presentation of educational institutions in Naberezhnye Chelny ;
- presentations of educational services and distribution of presentation materials and CD disks with information on educational services of higher education institution [5];

3 . Professional orientation projects:
- the intellectual project "Academic Marathon" (for pupils of 11 forms);
- the city competition "Future Leaders of Tatarstan" (for pupils of 10-11 forms);
- talk-show about problems of vocational guidance "My choice" (with participation of leading experts of the enterprises of the city, marketing specialists, sociologists, psychologists, representatives of mass media);
- festival of scientific and technical creativity of higher education institution "A step to perfection" (for pupils 9-11 forms);

4 . Preparation of professional orientation materials:
- preparation of an educational and methodical grant "Psychology of scientific and technical creativity";
- preparation of an educational and methodical grant "A complex of psychodiagnostic techniques for detection of tendencies of pupils and students to professions like "People – the Technician";

5 . Participation in conferences of a professional orientation profile:
- article preparation "Specifics of formation of professional intentions of the personality during joint projects "school -higher education institution"
  (V Republican scientific and methodical conference "The integrated system of professional education in conditions of a scientific and educational cluster: problems and ways of developments" ("NMK-KNITU (KHTI) — 2012"), on April 11, 2012. Kazan);
- Participation in the All-Russian scientific and practical conference with the international participation "New technologies of knowledge intensity of mechanical engineering: priorities of development and training" (on May 17, 2012);

6 . Participation in competitions and grants:

- Participation in 9 All-Russian open competition of teachers "Educational capacity of Russia" (An educational and methodical grant "A complex of psychodiagnostic techniques for detection of tendencies of pupils and students to professions like "People – the Technician") – 1 place;
- Participation in 1 Open All-Russian competition of author's programs, educational and methodical materials and virtual resources on development of creative activity of children, teenagers and youth" (An educational and methodical grant "Psychology of scientific and technical creativity");
- Project preparation for participation in competitive selection on receiving a grant of the Ministry of Education and Science of the Republic Tatarstan within implementation of the project "School After School" of Zakamye of Tatarstan). The grant on 1 million rubles is won.

For the purpose of the analysis of specifics of being trained students specialists psychologists of the Naberezhnye Chelny institute annually conduct researches of students of a first year on K. Rogers and R. Daymond's technique [6]. In 2012 414 first-year students of a full-time department took part in research.

Results of the conducted research are presented in tables and in figure 1.

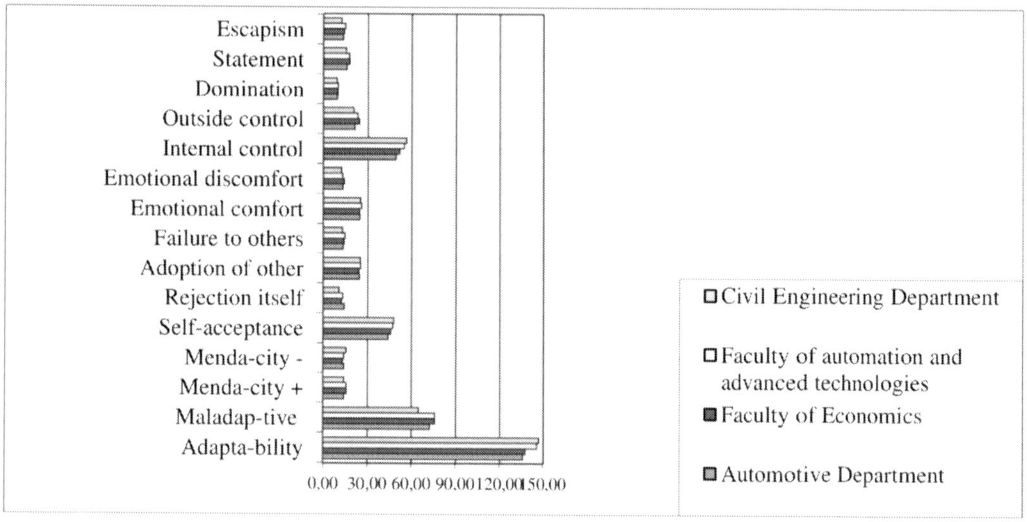

Fig. 1.

As a result of the conducted researches it was revealed that adaptability of students makes 141,3 points (at norm = from 68 to 136 points). Results of adaptability of students on faculties:
automotive faculty - 135,6 points;
economics department - 137,3 points;
automation faculty - 145,5 points;
construction faculty - 146,9 points.

The result is interpreted as the highest. Therefore, adaptability of students of faculties is at very high level; students of all faculties accept themselves and others, test emotional comfort [7]. In groups with the smallest number of points on a scale "adaptability" the corresponding psychological trainings [8] were provided. Thus, research of adaptation on K. Rogers and R. Daymond's technique showed that level of adaptation of pupils of a first year is at high level. Also anti-stressful training and training on motivation among masters of faculty of automation and progressive technologies was provided. Changes in system of professional education involve emergence everything new and new problems both personal, and the professional plan. Therefore monitoring of the relation of the personality to a profession, formation of the positive relation to future profession, desire to be realized in it demands search of new forms and innovations in vocational guidance in higher education institution [9]. The purpose of modern educational space - to allow each pupil and the student regardless of the

IOP Conf. Series: Materials Science and Engineering **69** (2014) 012028 doi:10.1088/1757-899X/69/1/012028

psychophysiological features, educational opportunities, tendencies to realize itself as the subject of own life, activity, communication [8]. One of the most important tasks of modern school is development in the identity of a creative initiative and independence, design and rationalization skills. In this regard the role of technical creativity in formation of the personality capable in the future to high-performance work, technically saturated production activity raises.

3.References

[1] . http://www.fcpro.ru/the Federal target development program of education for 2011-2015.
[2] . Kurbatskaya T B *Motivational structure of professional intentions in the course of option and vocational training in HIGHER EDUCATION INSTITUTION.* 1997 Abstract of the dissertation of candidate of psychological sciences. Kazan,.
[3] . Kurbatskaya T B *Shaykhelislamov R. F. Motivational structure of intentions of the personality.* 1999 Collection of Materials of the international scientific and practical conference. Naberezhnye Chelny
[4]. Tsagarelli E B *Strukture of system of psychology and pedagogical diagnostics and correction of the personality in educational process* 2012 Modern problems of science and education, No. 8,
[5]. Pryazhnikov N, Rumyantseva L. *Self-determination and vocational guidance of pupils.* 2013 M, Academy,
[6]. Raygorodsky D Ya *Practical psychodiagnostics. Techniques and tests. Education guidance.* 2001 Samara. Bakhrakh-M.
[7]. Bezyuleva G V *Psychology and pedagogical maintenance of professional adaptation of pupils and students.* 2008 M, MPSI.
[8] . Kurbatskaya T B *Psychology of work.* 2009 K., Yuniversum.
[9]. Tsagarelli E B *Structure of system of psychology and pedagogical diagnostics and correction of the personality in educational process* 2012 Modern problems of science and education, No. 8.
[10].Kurbatskaya T B, Shaykhelislamov R F *Research of motivational structure and features of communication of the personality* 1999 Scientific notes of the Kazan state university. Naberezhnye Chelny.
[11]. Petrushin S V *Psychologicall training in numerous group.* 2000 M, Akkad. Project.

Automatic control system of high precision welding of workpieces in mechanical engineering

I N Kuznetsov, V V Zvezdin, I H Israfilov, S M Portnov
Kazan Federal University,
423800, 68/19 Mira avenue, Naberezhnye Chelny, Russia

E-mail: chelny@kpfu.ru

Abstract. In this paper, based on the conducted patent research, the system of laser welding control with different geometry of weld and shapes of parts is developed. The method of monitoring the position of the spot of laser radiation in relation to the curved weld is worked out; it is based on the tracking the edges of the welded parts by low-power laser radiation reflected from the surface of the parts. It allows to make the positioning of the focus of laser radiation in relation to the juncture of the welded parts automatically.

1. Introduction

One of the trends of modern engineering is the widespread use of automated system of laser welding as one of the effective ways to reduce the costs of production by increasing the ratio of utilization of materials [1]. Provision of stability of properties and quality of weld, as well as high economic indicators, requires the application of automatic control system in the technological process (TP). Elaborated and currently used technological methods do not allow to solve fully all the problems of automation of the welding process in mechanical engineering.

2. Basic part

The quality of the obtained product has a complex dependence on the technological parameters of the process (radiation power density, speed of the movement of laser radiation, gaseous atmosphere, etc.) that causes the need to control the process by maintaining those parameters within the required limits. [2] This necessitates the use of new approaches to the control of the process of high-precision laser welding and improvement of automatic control system (ACS) TP.

To control the position of the focus of laser radiation (LR) in relation to the workpiece is possible using 4 - quadrant photo detector (PD) [3]. A sensitive layer of the multi-element coordinate photo detector (MCP) consists of several individual elements enclosed in a single housing. MPC application simplifies the construction of certain types of opto-electronic converters, as it excludes the mechanical scanning. Receiving information from the platforms of multi-element coordinate photo detector is performed with the help of high-speed commutators, which allow to determine the locus of the spot of LR in relation to the weld [4]. Fig. 1 shows the location of the spot of LR on the photosensitive platform.

IOP Conf. Series: Materials Science and Engineering **69** (2014) 012029 doi:10.1088/1757-899X/69/1/012029

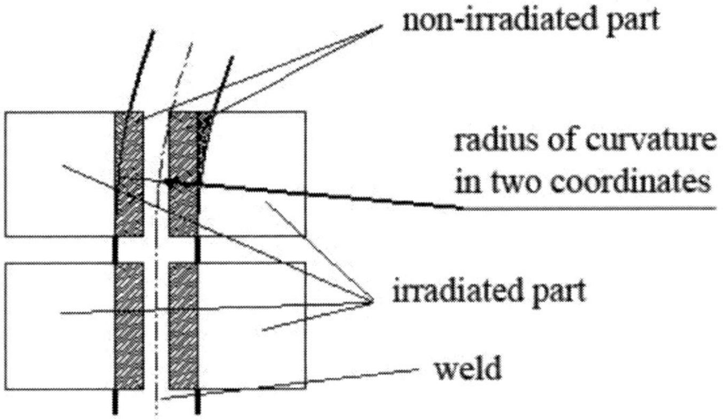

Fig. 1. The location of the light spot on the platforms of 4-quadrant photo detector.

When you change the curvature of the weld, there is a change in the ratio between the internal resistance of the illuminated and unilluminated surfaces of MCP platforms. This leads to a change in the differential values of resistance of platforms**.** Rotation of MCP to the angle $\Delta\varphi$ allows to balance these ratio, as well as to calculate the control signal to the drive of MPC by the worked out algorithm, taking into consideration the spatial positions of MCP and the optical system LTC [5].

Linearity region of the output signal Vx (x) and Vy (y) is defined by the inequalities:

$$x<l \quad \text{and} \quad y<l.$$

As it can be seen, when there is a quadratic light spot and parallelism of diagonals of the square to the axes x and y, the mutual influence of axes is absent.

The calculation of active resistance of the photosensitive layer of photo detector shows its dependence on the area of illumination:

$$R_{\Phi P} = \frac{\rho_н \rho_o l^2}{dl\left[(x-\Delta x)\rho_н + \Delta x \rho_o\right]} = \frac{\rho_н \rho_o l^2}{d\left[(S_п - S_н)\rho_н + S_н \rho_o\right]},$$

where **Rpc** - resistance of the photosensitive layer, ρN, ρO - resistivity in accordance with unlit and lit sites of photo detector;

l - length of the photosensitive platform;

SП, SH - area of the photosensitive layer, according to unlit and lit surfaces of photo detector.

As follows from the calculations, correction of MPC drive of the remote measurement of LR focus position on regard with the angle displacement between the optical axes of MCP and the tangent of the weld at the point of LR interaction with metal is defined by the following formula [6]:

$$\Delta\varphi = f\left(R_{\Phi Пп} - R_{\Phi Пн}\right),$$

where $\Delta\varphi$ is increment of the angle displacement between the optical axis of the photo detector and the tangent of the weld at the point of interaction of laser radiation with metal; RPDp, RPDn is a resistance of the photosensitive layer according to the complete and unlit surfaces of the photo detector.

However, the focal length of the laser backlight changes for the parts of complex shapes, which leads to an increase of the spot on the 4 - quadrant PD. To exclude this effect, it is necessary to create a parallel beam; it can be provided with the help of the collimator.

When submitting laser radiation on the weld, it's displacement in relation to the curved juncture and parts of complex shape is possible. To reduce the positioning error of LR, a feedback channel is introduced; it is based on the backlight of the juncture by the low-power laser and reception of the reflected radiation by the four-quadrant photo detector [7]. After processing on the corresponding units of ACS LTC, the error signal arrives to MCP, where the processing of received information according to the given algorithm takes place, and the error signal is issued through the NC on the appropriate drive. As the focal length of the low-power laser with a certain divergence of laser radiation will change according to the distance of the optical system to the part surface, it is necessary to apply the collimator, which converts the divergent beam of LR into a parallel one. Therefore, the diameter of the reflected radiation, which reaches a four-quadrant photo detector, remains practically unchanged, it reduces the positioning error of the beam of LTC. Correction of the positioning is carried out by the drives. Flowchart of the worked out ACS LTC is shown in Figure 2.

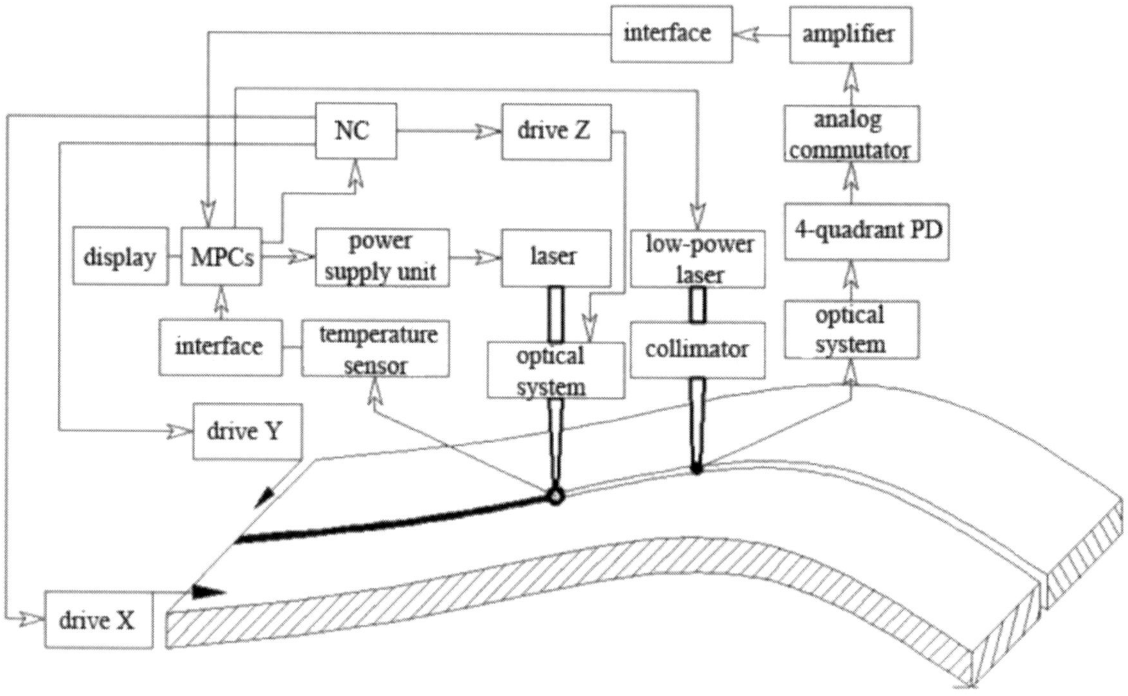

Fig. 2. Flowchart of ACS LTC of laser welding of metals.

The temperature in the processing zone is measured by the ratio of the absolute intensities by the formula [8]:

$$T = \frac{h\nu}{4\pi} \cdot \frac{Ag}{Z} \cdot Ni \exp\left(-\frac{E}{KT}\right),$$

where ν - frequency; E – power of drive; g- statistical weight of level; A - transition probability; h-Planck constant; k-Boltzmann constant; T - temperature; Z - statistical sum; Ni-ion concentration.

When using this formula, the temperature of the investigated zone of the source of spectrum of radiation is estimated [9], it determined a definite value of the ratio of line intensities I1 / I2, according to which it is equal to:

$$T_0 = \frac{E_1 - E_2}{\left(\ln\frac{I_1}{I_2} - \ln\frac{\lambda_2 g_1 A_1}{\lambda_1 g_2 A_2}\right)k}.$$

Using the known PD, providing the combination of plane of analysis of polarizer of one of the photosensors with the plane of polarization of the analyzed radiation [10], it is possible, by measuring the voltage of photosignal from the sensors, to determine the degree of linear polarization of the incident radiation by the formula:

$$P = \frac{S_1}{S_0} = \frac{J_{\psi=0} - J_{\psi=90}}{J_{\psi=0} + J_{\psi=90}} = \frac{U_{\psi=0} - U_{\psi=90}}{U_{\psi=0} + U_{\psi=90}},$$

using the well known PD, one can determine the degree of linear polarization of radiation, azimuth of which coincides with the plane of the analysis of one of the polarizers of PD.

Measuring voltage of photosignal, we can calculate the degree of polarization P and azimuth of polarization ψ of the analyzed radiation:

$$P = \frac{\sqrt{S_1^2 + S_2^2}}{S_0} = \frac{\sqrt{\left(U_{\psi=0} - U_{\psi=90}\right)^2 + \left(U_{\psi=45} - U_{\psi=-45}\right)^2}}{U_{\psi=0} + U_{\psi=90}},$$

$$\psi = \frac{1}{2} arctg \frac{S_2}{S_1} = \frac{1}{2} arctg \frac{U_{\psi=45} - U_{\psi=-45}}{U_{\psi=0} - U_{\psi=90}},$$

3.Conclusions

The method of monitoring the position of the spot of laser radiation in relation to the curved weld is worked out; it is based on the tracking of the edges of welded parts by low-power laser radiation reflected from the surface of parts. This allows to conduct the positioning of the focus of laser radiation in relation to the junction of the welded parts automatically. As a result, it reduces the impact of vibration of the welded parts and the laser head on quality indicators of TP.

3. References

[1] *Technological processes of laser processing: Textbooks for higher education* / I N Shiganov, A I Misyurov; under the editorship of A.G. Grigoryants; - Moscow: MSTU named after N.E. Bauman, 2006. - 664 p

[2] Vedenov A A, Gladush G G 1985 *Physical processes in laser processing of materials.* – M.: Energoatomizdat

[3] Patent № 2193168, Russia , IPC G01J 4/00 *Method of measuring the degree of polarization* / Ziyatdinov R R, Zvezdin V V, Gumerov A F, Sabirov I S Application № 2000118517 /28

[4] *Positioning of the laser radiation in relation to a weld as a quality indicator of the process* / V V Zvezdin, A V Khamadeev, R G Zagiriv, I R Shangaraev // Vestnik KSTU named after A.N. Tupolev, № 3, 2008. - S 17-19

[5] *Calculation of the parameters of the polarization-sensitive multispectral photodetector for measuring the true temperature.* / V V Zvezdin, V V Zamorsky, R F Zaripov, V S Karimov // Information and socio-economic aspects of modern technology. - Interactive electronic scientific and technical journal, 2003 № 9 P 10- 17

[6] *Improving the quality of welded joints of nodes and car parts in laser welding* / V V Zvezdin, V V Zamorsky, V S Karimov, I S Sabirov // Problems of quality and operation of vehicles. Materials III International Scientific and Technical Conference P 1. Penza - PSUAC, 2004 S 54-58

[7] *Calculation of the control channel by the position of the laser radiation during the process of welding* / V V Zvezdin, R M Aleev, A V Khamadeev / Designing and investigation of technical systems: Interuniversity scientific collection. № 11. - Nab. Chelny Univ. INEKA, 2008 P 33-38
[8] Azzam R, Bashara N 1981 *Ellipsometry and polarized light.* Moscow: World 583 p
[9] Svet D Y 1982 *Optical methods for measuring the true temperature*. Moscow: Nauka 296 p
[10] *Polarization of the thermal radiation as informative parameter of the interaction of laser radiation with metals* / V V Zvezdin, A V Khamadeev, S M Portno, R F Zaripov // Vestnik KSTU named after A.N .Tupolev, № 3, 2008 P 88-91

Magnetron discharge volt-ampere characteristic investigation at thin film coating process

N F Kashapov, A G Luchkin and G S Luchkin
Kazan (Volga Region) Federal University, 18 Kremljovskaja str., Kazan 420008, Russian Federation

E-mail: AGLuchkin@kpfu.ru

Abstract. Magnetron discharge at reactive and working gases mixture atmosphere current-voltage characteristic (*I-U*) for different sputtering parameters is investigated. It is shown, that form of volt-ampere characteristic doesn't depend on gas supply scheme at vacuum chamber pressure 4-$6 \cdot 10^{-2}$ Pa. Reactive gas (oxygen) flow increasing leads to making *I-U* transition part wider and amplification of difference between top and bottom parts of hysteresis loop *I-U*. Discharge voltage is less at oxygen and argon gases mixture atmosphere than at argon atmosphere.

1. Introduction

It is well known that magnetron sputtering is applied for making functional coatings including biocompatible on different materials [1-4]. Example of biocompatible coatings are thin films of titanium and it's compounds (TiO_x, TiN, etc.) [5,6]. TiO_x coatings can be made by titanium oxide sputtering at argon atmosphere or by titanium sputtering at oxygen and argon gases mixture atmosphere. During metals magnetron sputtering at oxygen and argon gases mixture atmosphere on metal target surface appears oxide thin film [7], and volt-ampere characteristic has hysteresis shape [8,9].

Experimentally determined that depending on target surface oxide covering degree there are several sputtering regimes. If target surface totally covered by oxide film, metal oxide parts are sputtered. If target surface covered partly – metal parts. Transition between regimes is uncontrolled. Difference between regimes is well shown at shape of *I-U* – it is hysteresis. There some methods to avoid hysteresis [10]. Also there is opportunity to use both regimes.

I-U hysteresis has metallic, transition and oxide parts. For sputtering at both regimes it is necessary to investigate transition part of *I-U* hysteresis. The article aim is to determine discharge parameters influence on transition part of *I-U* hysteresis.

2. Experimental plant and methods

Functional schema of experimental plant is showed at Pic. 1. It consists of: Vacuum chamber 1, door 2, pump system 3, two magnetrons 4, power supply of magnetron and source of ions 5, system of gas input, system of photometric control of film thickness that include spectrophometer S-100, moving system 8 of substrates 9, heating system 10, windows 11, valve for atmosphere air 12, filled-system transducer 13, cooling system 14, ion source 15 [11,12].

There is control stand for all systems. Pumping system includes mechanical pump 2NVR90B and diffusion pump ND-200 and provides Vacuum at chamber about 10^{-3} Pa. Pressure registration system consists of vacuum-gauges VT1-4 (thermocouple method) and VMB1-2

IOP Conf. Series: Materials Science and Engineering **69** (2014) 012030 doi:10.1088/1757-899X/69/1/012030

(ionization method). Gas input in chamber realized by leak valves VTT02.01.3T.000 and controller RRG-8. Gas flow counted by constant volume method [11].

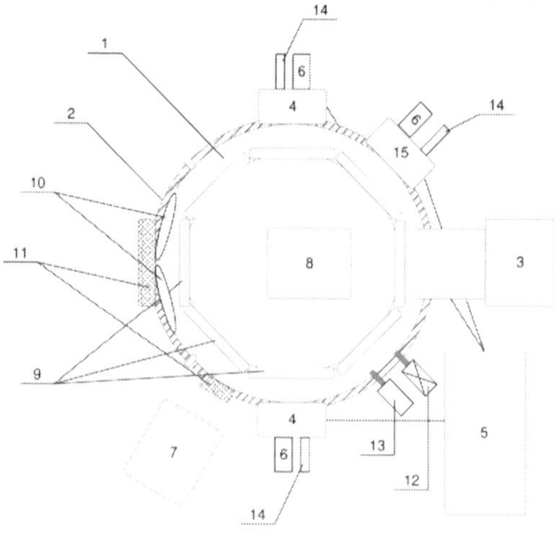

Pic. 1 Functional schema of experimental plant: Vacuum chamber 1, door 2, pump system 3, two magnetrons 4, power supply of magnetron and source of ions 5, system of gas input, system of photometric control of film thickness that include spectrophometer S-100, moving system 8 of substrates 9, heating system 10, windows 11, valve for atmosphere air 12, filled-system transducer 13, cooling system 14, ion source 15.

Current-voltage characteristic measured for different reactive and working gases ratios. Plant permits two cases of gas supply: into target region (schema “1”) and into substrate region (substrate “2”). Discharge parameters were: voltage from 100 to 700 V, current density from 4 to 24 mA/sm^2, chamber pressure 0,1 – 0,8 Pa. Target material – titanium, argon flow 0,4 – 1,2 mg/sec, oxygen flow was 0,2 – 0,7 mg/sec.

3. Results and discussion

Discharge *I-U* characteristics at oxygen and argon atmosphere (gas supply schema “1”) are shown at Pic. 2. Oxygen flow into vacuum chamber was 0,4 и 0,7 mg/s. Pressure in vacuum chamber was $4\text{-}6 \cdot 10^{-2}$ Pa. Current density was from 4 to 20 mA/sm^2, voltage increased from 200 to 600 V.

At increasing current density of discharge at oxygen and argon atmosphere (“O_2+Ar ↑”) voltage was less than in case of discharge at argon atmosphere (“Ar ↑↓”). Transparent oxide films was coated at this *I-U* region, because oxygen presence caused oxide film on magnetron target surface and discharge power was not strong enough to sputter it. This regime ended at current density about 18 mA/sm^2, when high discharge power caused full sputtering of oxide film on magnetron target surface. Than began metal (elemental) regime (metal film sputtered). Metal regime voltage deviation from sputtering at argon atmosphere was insignificant. Therefore were identical processes. At decreasing current density (“O_2+Ar ↓”) was reverse transition at 8 mA/sm^2.

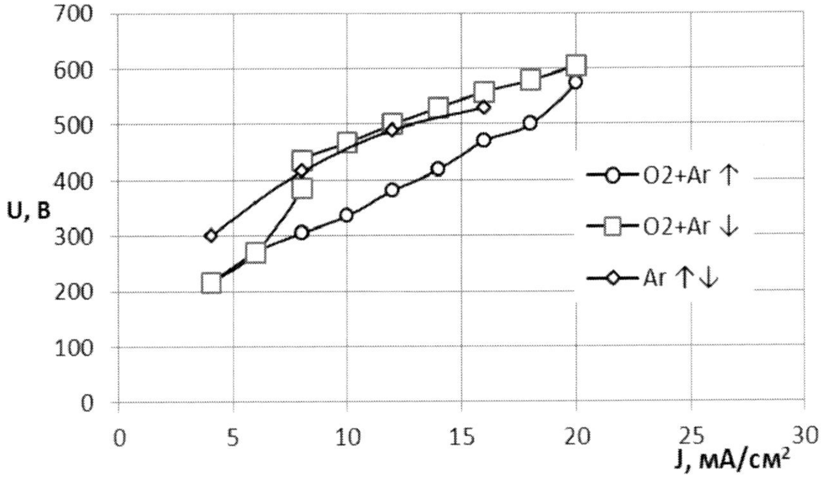

a)

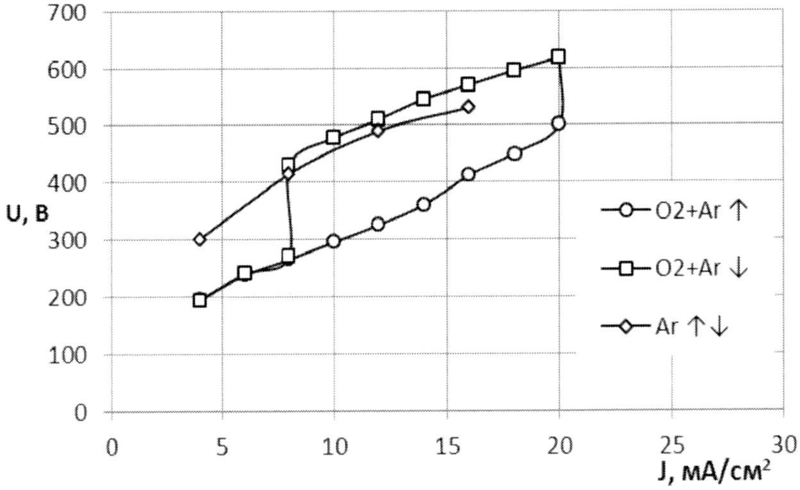

b)

Pic. 2 *I-U* for gas supply schema «1»:
a), b) – argon (flow 0,4 mg/s) and oxygen (flow 0,4 and 0,7 mg/s respectively) mixture

Oxygen flow increasing (Pic. 2, b) didn't change qualitatively *I-U* form, but increased difference between metal and oxide regimes and transition region. Transition to metal regime was at 20 mA/sm^2. Voltage of oxide regime was lower 500 V.

Discharge *I-U* characteristics at oxygen and argon atmosphere (gas supply schema "2") are shown at Pic. 3. Oxygen flow into vacuum chamber was 0,4 и 0,7 mg/s. Pressure in vacuum chamber was 4-6·10^{-2} Pa. Current density was from 4 to 24 mA/sm^2, voltage increased from 200 to 680 V.

Oxygen flow leads to lower voltage (30-50% less) at current increasing and to hysteresis appearing (Pic. 3). At discharge current increasing oxide deposits. Transition to metal regime was at current density 12 mA/sm^2. Reverse transition to oxide coating deposition regime at decreasing current density was at 6 mA/sm^2.

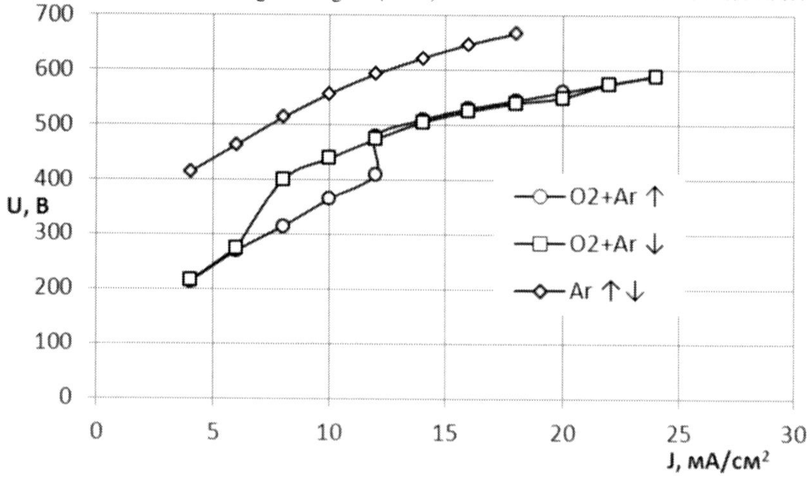

a)

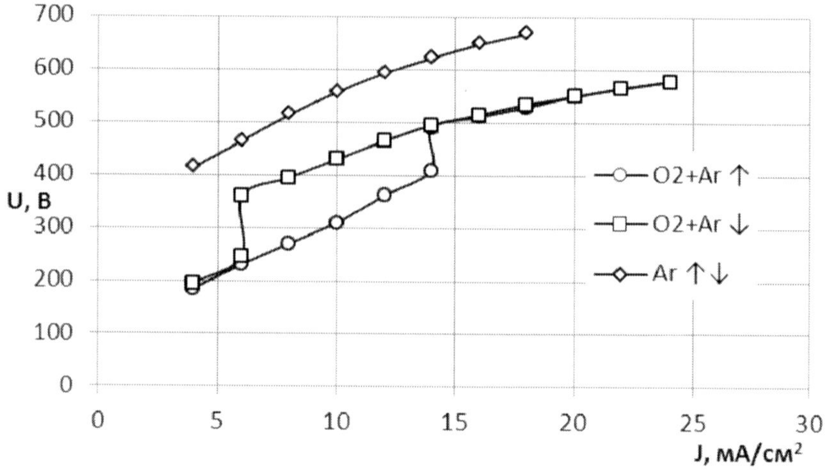

b)

Pic. 3 *I-U* for gas supply schema «2»:
a), b) – argon (flow 0,4 mg/s) and oxygen (flow 0,4 and 0,7 mg/s respectively) mixture

Oxygen flow increasing in 2 times (Pic. 3 b) didn't change I-U from qualitatively, but voltage difference between metal and oxide regimes enlarged. Transition state also increased. Transition to metal regime was at 14 mA/sm^2. Voltage of oxide regime was lower than 400 V.

Therefore reactive gas (oxygen) flow rate significantly effects on appearing I-U hysteresis loop and on metal magnetron sputtering ability at presence of reactive gas in vacuum chamber.

4. Conclusions

Magnetron discharge at reactive and working gases mixture atmosphere *I-U* characteristic for different sputtering parameters is investigated. It is shown, that form of volt-ampere characteristic doesn't depend on gas supply scheme at vacuum chamber pressure 4-6·10^{-2} Pa. Reactive gas (oxygen) flow increasing leads to making *I-U* transition part wider and amplification of difference between top and bottom parts of hysteresis loop *I-U*. Discharge voltage is less at oxygen and argon gases mixture atmosphere than at argon atmosphere.

References

[1] Galyautdinov R.T., Kashapov N.F. and Luchkin A.G. 2014 *Remont, Vosstanovlenie, Modernizatsiya* vol. 1 pp. 37-40

[2] Kashapov N.F., Luchkin A.G. and Luchkin G. S. 2010 *Bulletin of KGTU* vol. 2 pp. 340-345
[3] Galyautdinov R.T., Kashapov N.F. and Luchkin G. S. 2005 Prikladnaya Fizika vol. 6 pp. 88-93
[4] Galyautdinov R.T., Kashapov N.F. and Luchkin G. S. 2003 *Welding International* vol. 17 pp. 655–658
[5] S. Mandl and B. Rauschenbach 2002 *Surf. Coat. Technol* **156** 276
[6] C.-M. Lin and S.-K. Yen 2004 *J. Electrochem. Soc.* **151** D127
[7] Abe et al. 2005 *J. Vac. Sci. Technol.* A **23**(5)
[8] Komlev R.A., Shapovalov V.I. and Shutova N.S. 2012 *Technical physics letters* vol. 82 pp. 134-136
[9] Galyautdinov R.T., Kashapov N.F., Luchkin A.G. and Luchkin G. S. 2013 *11 International conference of a "Films and Coatings - 2013"* (*Saint Petersburg*) pp. 279-281
[10] Dostanko A.P. 2009 *Technological processes and systems at microelectronics: plasma, electron-ion-beam, ultrasonic* (Minsk: Bestprint)
[11] Galyautdinov R.T., Luchkin A.G. and Luchkin G. S. 2013 *IV Republican scientific-technical conference Low-temperature plasma in processes of deposition of functional coatings* (*Kazan*) pp. 155-161
[12] Kashapov N.F., Luchkin A.G. and Luchkin G. S. 2013 *J. Phys.: Conf. Series* **479** 012019

IOP Conf. Series: Materials Science and Engineering **69** (2014) 012031 doi:10.1088/1757-899X/69/1/012031

Unpowered stand for testing of car diesel engine turbochargers

V A Luscheko, V N Nikishin
Kazan Federal University,
423800, 68/19 Mira avenue, Naberezhnye Chelny, Russia

E-mail: chelny@kpfu.ru

Abstract. The non-motor test stand for turbocharger is represented in this article.

1. Introduction

The internal combustion engine is the most popular source of energy. Modern state of drive engineering is characterized by the production of highly accelerated engines. It is necessary to ensure a high level of quality and safety standards. Developers have to solve the complex problems of quality, reliability and improve the economic and environmental performance of diesel. It is known that 60% of all claims that arise during the guarantee period are due to the imperfection of development. Product quality as a set of properties is laid in the process of research, design and technological elaborations, created in the production process and manifested in the operation. Based on this, the most important global challenge of quality should be provided at the stage of product design and technology [1; 2]. The authors created the unpowered stand, working on open circuit, designed for testing of automotive diesel turbochargers (TCR) to form their quality (efficiency and reliability) during the design and operational development. This stand allows you to define the characteristics of compressors and turbines [3; 4; 5; 6; 7; 8, 9], and conduct experimental study of bearing assemblies TCR [10; 11].

.

2. Basic part

The feature of the stand is the automated control system, through which all systems and units of the stand are controlled by a personal computer, which not only improve the accuracy and the quality of ongoing trials, to reduce measurement error, but also cuts the time required, that is especially important in the mass production .

Schematic diagram of the non-motorized of the stand is shown in Fig. 1.

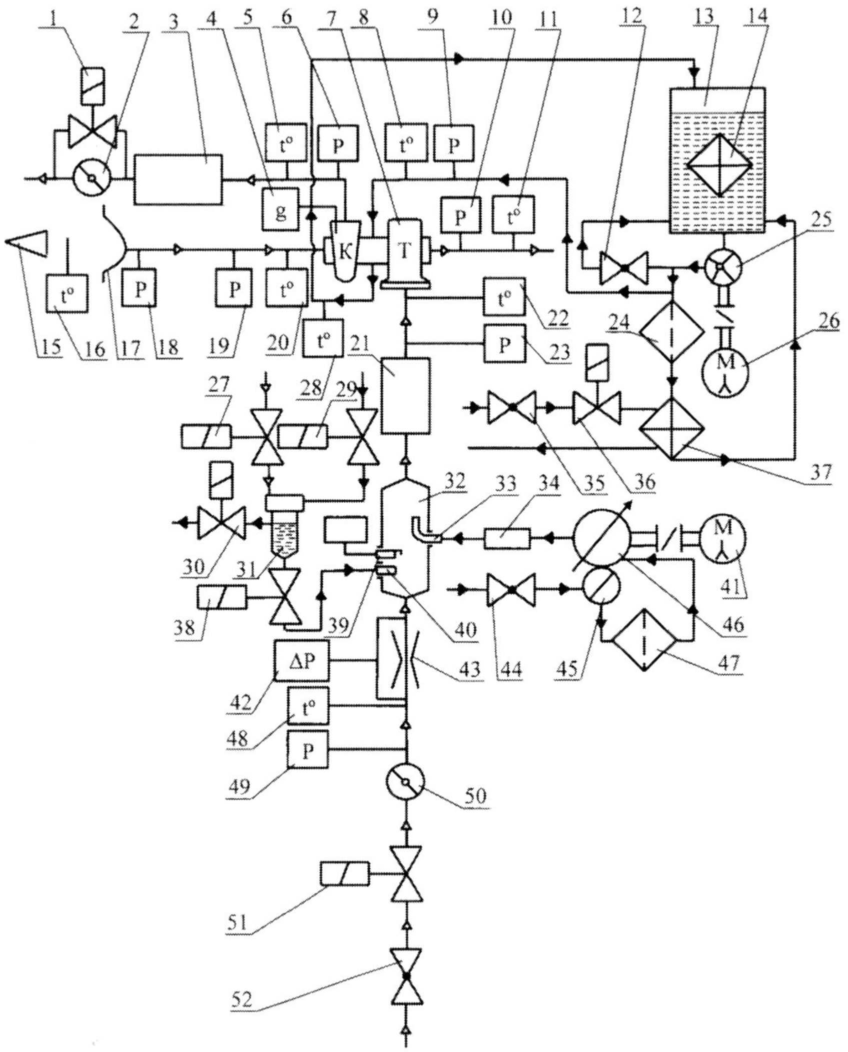

Fig. 1. Schematic diagram of the non-motorized stand:
1 - bypass; 2 - throttle the compressor; 3 - receiver; 4 - vibration sensor; 5, 8, 11, 16, 20, 22, 28, 48 - the temperature sensors; 6, 9, 10, 18, 19, 23, 42, 49 - the pressure sensors; 7 - experienced TCR; 12 - Oil flow regulator; 13 - oil tank; 14 - oil heater; 15 - Sensor frequency drive rotation of the rotor; 17 - nozzle flow meter of the compressor; 21 - Receiver; 24 - oil filter; 25 - pump oil; 26 - motor oil pump; 27, 29, 30, 36, 38 - valve; 31 - starting volume; 32 - combustion chamber; 33 - Working nozzle; 34 - Receiver; 35 - tap water supply; 37 - oil-water heat exchanger; 39 - the ignition switch to the spark plugs; 40 - starting atomizer; 41 - electric fuel pump; 43 - turbine nozzle flow meter; 44 - tap the supply of fuel; 45 - priming pump fuel pump; 46 - Injection; 47 - fuel filter; 50 - throttle before the turbine; 51 - shut-off valve; 52 - air supply valve from the original system.

To control a stand there is a a program that allows you to: measure the current settings; specify the required values of the frequency of the rotor shaft of the TCR and the gas temperature at the turbine inlet and manage them; store in the database and Excel file information on the subject of the TCR, the test program and test results; production led calibrate and adjust the dimension of the measured parameters.

The scheme of data collection and control how stand for testing of TCR is shown in Fig. 2.

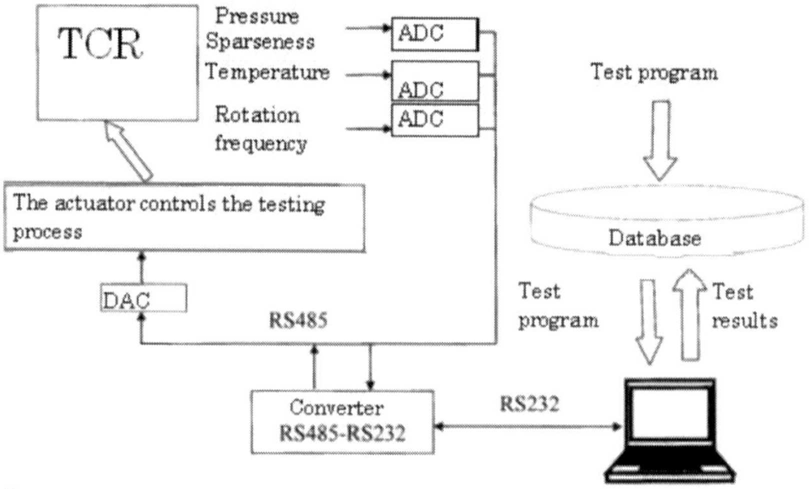

Fig. 2. The scheme of data collection and to control operation of the stand for-Test of TCR

The scheme of data collection and to control operation of the stand consists of the following elements:

- PC management program;
- Analog-to-digital converters (ADC) connected to the network and RS-485 for measuring the sensor signals, analog output devices, as well as to manage processes that occur during the test of TCR;
- Temperature sensors, pressure and speed, the signals from which ADC are measured;
- Digital-to-analog converters (DACs) and executive mechanisms for process control tests.

In order to conduct research tests collection of additional information from the transducers of temperature and pressure (redundant inputs) is provided. The program of experiments, collection and processing of data in this case undergoes a corresponding change.

3. References

[1] *Establishing and maintaining the quality of automotive diesel*. Part I / V Nikishin; Ministry of Education and Science; Kama state Engineering-econ. Acad. - Nab. Chelny: Publishing House of the Kama st. Engineer-Econ. Acad., 2006 456 p 428-452 p (Monographic studies: technique)

[2] *Establishing and maintaining the quality of automotive diesel*. Part II / V Nikishin dts; SEE "Kama state Engineer-Econ. Acad. ". - Naberezhnye Chelny Publishing House of the Kama st. Engineer-Econ. Acad., 2008.175 p - pp 149-173 (Monographic studies: technique)

[3] Zinner K 1978 *Supercharging of internal combustion engines = Aufladung-von Verbrennung* - gasmotoren: Translated from the German / Ed. Dr. tehn. on-yk N.N.Ivanchenko. - L., Mechanical Engineering. Leningrad. Dep. 264 p with illustration

[4] GOST(state standard) R 53637-2009. Motor and turbochargers. General technical requirements and test methods

[5] Yurenkov V N 2003 *Basic principles and research compressors and turbines for turbochargers diesel motorless stand* / Polzunovskii Gazette № 1-2 P 70-77

[6] Stechkin B S, Kazandzhan P K, Alekseev L P, Govorov A N Nechaev, Y N, Fedorov R M 1956 *Theory of jet engines. Turbomachinery*. Ed. Academician BS Stechkin. M. Oborongiz 548 p with illust.

[7] Naundorf D 2008 *Turbocharger test stand with a hot gas generator for high-performance supercharging systems* MTZ Vol. 69 p 22-24

[8] Shaaban S 2004 *Experimental investigation and extended simulation of turbocharger non-adiabatic performance* Hannover 228 p

[9] Galyautdinov R T and Kashapov N F 2001 Iz. VUZ. *Aviatsionnaya Tekhnika* **2** 80.

[10] *High-speed turbocharged diesel engines*. M., "Engineering", 1976. 288 p Auth. A E Simson, V N Kaminsky, Y B Morgulis, G M Povetkin, A B Azbel, V A Kochetov

[11] Hanin N S, Aboltine E V, Lyamtsev B F 1991 *Car and other turbocharged engines*. - M.: Mechanical Engineering 336 p with illust

Modified cationic membranes for water purification, and their selective permeability

G V Mavrin, D D Fasullin
Kazan Federal University,
423800, 68/19 Mira avenue, Naberezhnye Chelny, Russia
R G Melkolvan
Moscow state mining University

E-mail: denr3@yandex.ru

Abstract. Wastewater containing heavy metal ions pose a significant toxicological risk to aquatic ecosystems and humans. The common problem of modern engineering technology is the development of environmentally friendly systems with a closed-circuit and a minimum waste. The ion exchange membrane can significantly reduce the cost of wastewater treatment and provide a high degree of purification.

1. Introduction

One of the main categories of wastewater at the enterprises of machine-building industry are the galvanic wastewater with high content of ions of metals.

Membrane separation methods, in particular the reverse osmosis (RO) are ones of the most promising, allowing to significantly reduce the costs of wastewater treatment and water of any required quality [1; 2]. But in baromembrane processes in particular in reverse osmosis high operating pressure of 4 - 5 bar for waters with mineralization up to 2 g/l is required. Research of influence of salt concentrations on the capacity of membranes was conducted on the membrane reverse osmosis EMO-N 45-300. The research was conducted on the laboratory unit membrane separation [3].

1.1. Back osmotic separation solutions

With increasing concentration of sodium chloride in a partial solution permeability and selectivity of the membranes have deteriorated. Curves characterizing the dependence of selectivity membrane concentration of sodium chloride are given on Fig.1.

According to research results, it can be concluded that the reverse osmosis at low concentration of the electrolyte is ineffective. The region of values of concentration of the electrolyte in the initial solution of 0.1 g/l to 2 g/l is an optimal selective permeability for reverse osmosis for separation of the solutions.

1.2 Ion-selective membrane

Ion-selective membranes possess a high degree of separation solutions [4], with both low and high concentration of dissolved salts, as well division of solutions by the use of ion-selective membranes occurs at low operating pressures of 0.1 to 2 bar, depending on the substrate, into which a working layer is applied.

2. Methods

In this paper an ion-selective membrane was obtained, working layer of which is polyaniline (PANI). PANI (Fig. 2) attracts the attention of researchers due to their optical and electrochemical properties, high chemical stability and increased selectivity of migration [5; 6; 7].

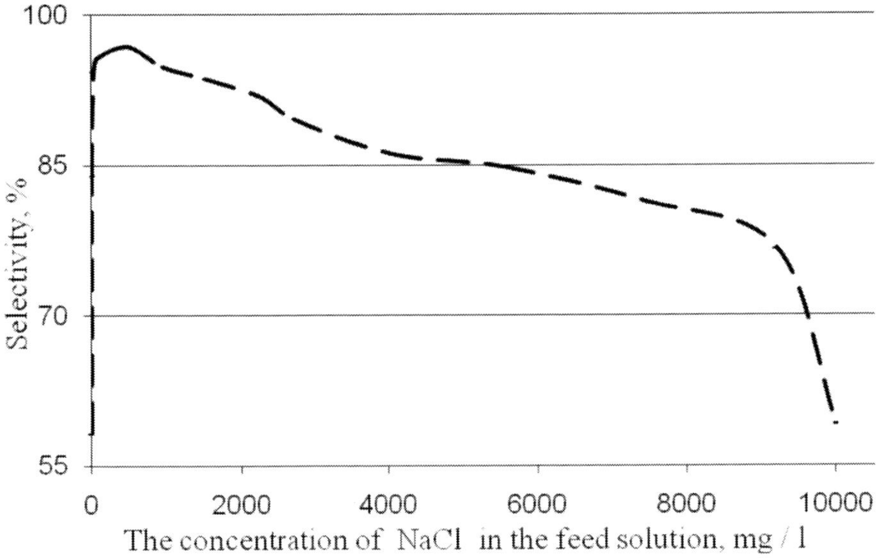

Figure 1. The dependence of selectivity of membrane from the concentration of dissolved substances (NaCl) in an initial solution.

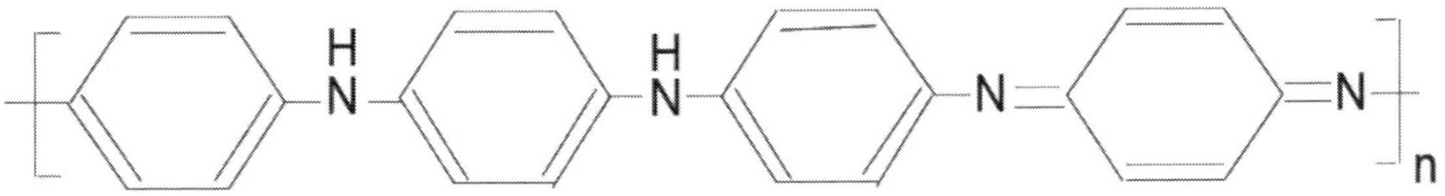

Figure 2. Structure of polyaniline

In the present work as a matrix for the polymerization of aniline PTFE membranes and nylon with pore dimensions 0.45 microns were used. The modification of the membranes to form on the surface and in the pores of the layer PANI, which is cationic [8], and allows to receive the ion-exchange membranes, not inferior to the selectivity of the reverse osmosis on a number of cations.

The synthesis of membrane surface PANY distribution was carried out by polymerization of aniline directly into the matrix of the membranes. Particles of PANY were formed directly in the matrix membrane, as evidenced polymer color changing to dark green. The change in the structure of membranes recorded with the help of a microscope, the image of the structure of the original and modified membranes is presented in figure 3.

3. Results and discussion

The pictures show that the introduction of ammonium persulfate in the membrane aniline saturated (method 2), immediately leads to the flow of oxidation, polymerization, as the concentration of aniline high in it.

The capacity of membranes was determined to pass through the membranes of certain volume of distilled water. The results are represented in table 1.

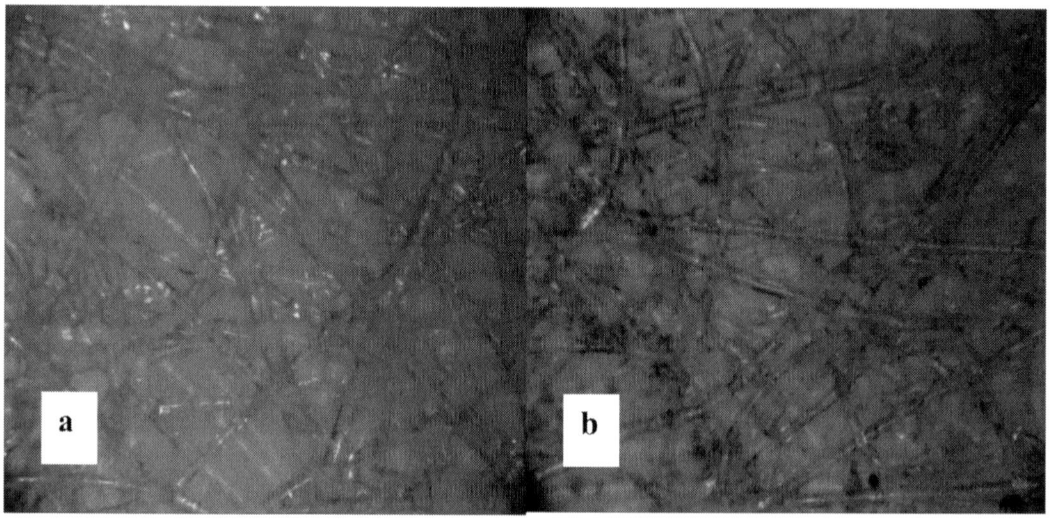

Figure 3. Structure of PTFE membrane: a) the initial PTFE membrane, b) PTFE membrane modified by PANI layer by method 2 (zoomed 200 times)

Table 1. Average permeability of membranes for distilled water

Membrane	Permeability, l/hour
PTFE *	2,29
PTFE +PANI (1 method)*	2,21
PTFE + PANI (2 method)*	1,93
Nylon	5,04
Nylon + PANI (1 method)	4,93
Nylon + PANI (2 method)	4,76

* before filtration membrane was spacewalked by acetone

It is known that the cation-exchange membrane is capable in a greater degree to adsorb cations than anions [9]. To study the selectivity of the original and modified membranes, iron chloride(III), zinc sulfate(II) and chromium ions (GSO), copper sulfate, Nickel ions (GSO), the ions of manganese (GSO), cobalt ions (GSO) solutions were let through membranes. The results are presented in table 2.

The selectivity of the membranes was calculated by the formula:

$$\varphi = (Cf - Cp) / Cf, \quad (1)$$

where Cf is the concentration of the solute in the initial solution and Cp is the concentration of dissolved substances in the filtrate[10].

Table 2. Selectivity nylon membranes with working PANI layer

No.	Indicator	The concentration of ions, mg/l		Selectivity, %
		source	after cleaning by membrane*	
1	Fe	5,57	0,015	99,3
2	Fe*	5,57	0,040	99,7
3	Cu	10,3	0,092	99,1
4	Mn	16,5	5,11	69,0
5	Cr	1,19	0,54	48,9
6	Cr*	1,19	0,62	55,0
7	Ni	316	30,3	90,4
8	Zn	8,22	3,38	59,0
9	Zn*	8,22	3,95	51,6

*nylon membrane with a working layer MRS obtained by method 1

4. Conclusions.

According to the results we can conclude that the original microfiltration membranes with pore size 0.47 microns have no selectivity on ions of metals. Modified polyaniline membranes have high selectivity and permeability to remove metal ions compared to reverse osmosis membranes.

References

[1] Sudilovskiy, P.S. Razrabotka sovmeschennogo flotatsionno-membrannogo protsessa ochistki stochnyih vod ot tyazhelyih metallov: avtoreferat dissertatsii kand. tehn. nauk. Moskva, 2007.

[2] Dyitnerskiy Yu. I. Baromembrannyie protsessyi. Teoriya i raschet. – M. Himiya, 1986. – 272 s.

[3] Fazullin D.D., Mavrin G.V., Melkonyan R.G. «Faktoryi, vliyayuschie na baromembrannyie protsessyi razdeleniya vodnyih rastvorov», Ekologiya promyishlennogo proizvodstva: Mezhotr. nauch.-prakt. zhurnal/FGUP «VIMI», 2012. Vyip.4.S. 65 – 70.

[4] Nagarale RK, Gohil GS, Shahi VK, Trivedi GS, Rangarajan R., 2004. Preparation and electrochemical characterization of cation- and anion- exchange/polyaniline composite membranes. Journal Colloid Interface Sci., 277(1): 162-171. Data obrascheniya: 10.01.2014, www.ncbi.nlm.nih.gov/pubmed/15276053.

[5] Ivanov V.F., Gribkova O.L., Cheberyako K.V.,Nekrasov A.A., Tverskoy V.A., Vannikov A.V. // Elektrohimiya. 2004. T. 40. # 3. S. 339.

[6] Gosh M., Barman A., Meikap A.K. et. Al. // Phys. Lett. A. 199. V.260. P. 138.

[7] Tan S., Laforgue. A, Bélanger D., 2003. CHARACTERIZATION OF SULFONATED CATION-EXCHANGE MEMBRANES MODIFIED WITH POLYANILINE. Fuel Chemistry Division Preprints 48(1): 386. Дата обращения: 23.11.2013,web.anl.gov/PCS/acsfuel/preprint%20archive/Files/48_1_New%20Orleans__03-03_0530.pdf

[8] Asif A., Margaret M., Darrell A., 2011. Surface and Charge Transport Characterization of Polyaniline-Cellulose Acetate Composite Membranes. J. Phys. Chem. B, 115 (7): 1652–1661. Дата обращения: 23.11.2013, pubs.acs.org/doi/ipdf/10.1021/jp109455m.

[9] Lyisova A.A., Stenina I.A., Dolgopolov S.V., GorbunovaYu.G., Kononenko N.A., Yaroslavtsev A.B.// VYiSOKOMOLEKULYaRNYiE SOEDINENIYa, Seriya B, 2011, tom 53, # 1, s. 130–136.

[10] Mulder M. Vvedenie v membrannuyu tehnologiyu: Per. s angl. – M.: Mir, 1999. – 513 s.

Innovative Mechanical Engineering Technologies, Equipment and Materials-2013 IOP Publishing
IOP Conf. Series: Materials Science and Engineering **69** (2014) 012033 doi:10.1088/1757-899X/69/1/012033

Practical experience of using thermal-mass flowmeters at the registration associated (free) petroleum gas

M G Fazlyyyakhmatov, N F Kashapov, Kh A Khayritonov, D K Lazarev and V K Lazarev
Kazan (Volga Region) Federal University, 18 Kremlyovskaya str., Kazan 420008, Russian Federation

Email: mfazlyjy@kpfu.ru

Abstract. The results of field tests of thermal-mass flowmeter TurboFlow TFG-S in comparison with ultrasonic flowmeter Dymetic-1223K at existing oil and gas extraction object are given in the article. Measured medium – associated (free) petroleum gas.

1. Introduction

The general trend of increasing volumes of hydrocarbon production in the world today caused by the increasing needs of the industry. For a more rational use of the products extraction and reducing polluting the biosphere emissions by burning in the Russian Federation developed and released a number of documents.

In particular, Resolution №1148 of 11.08.2012, "On peculiarities of calculating fees for emissions ...", and the Order of the President of the Russian Federation № PR-1461 from 10.08.2007 on the rational use of associated petroleum gas (APG). They envisage:

- Maximum permissible limits burning in torch and (or) dissipation - no more than 5% of the volume extraction APG;
- Bringing usage (utilization) of APG to 95% of total extraction;
- Introduction of compulsory monthly report for the extraction and collection of APG;
- Equipment of the oil and gas production objects of APG measuring instruments (MI), in the absence of MI, which confirm the actual volume extraction, use and burning - a multiple increase penal sanctions;
- In the case of violations of the established requirements for the organization of accounting APG - recall subsoil licenses.

GOST R 8.733-2011 [1] (was put into effect 1.03.2012) is the main regulatory instrument for the organization of APG accounting, which lists the terminology, classification systems measure the amount of gas, requirements to systems measure the amount of gas and it metrological assurance.

A feature of the APG is the physical manifestation of its component composition in the measured pipeline under different conditions.

Accounting for such gas is complicated by the fact that the hydrocarbons and water droplets flying in the gas pipe are deposited on the sensitive elements of flowmeter accumulate in the pipe and are transformed into a stream of fluid, whereby indications of counters are not always reliable.

2. Equipment and Tools

There are basically two different methods for determining the flow rate (volume) of the gas: mass (obtain immediately under standard conditions) and volumetric (obtain flow rate (volume) in the working (current) conditions, which then need to be brought to standard conditions).

The former include thermal-mass and Coriolis flowmeters. The second - turbine, rotary, ultrasonic, vortex and optical.

In this article the results of field tests of thermal-mass flowmeter TurboFlow TFG-S in comparison with ultrasonic flowmeter Dymetic-1223K at existing oil and gas extraction object are given, according to [1].

Ultrasonic flowmeter Dymetic 1223K (Figure 1a) uses correlation method for determining the flow rate (Figure 1b).

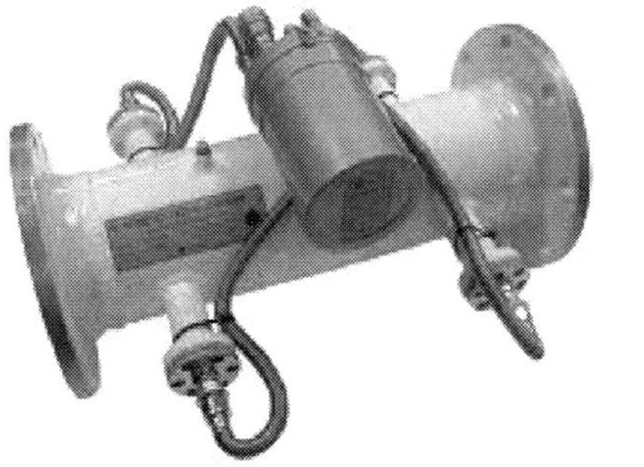

Figure 1a. Exterior view of the ultrasonic flowmeter Dymetic 1223K.

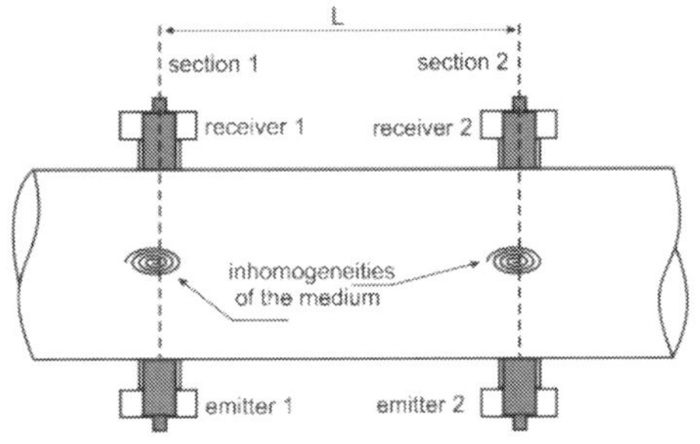

Figure 5b. The operating principle of the ultrasonic flowmeter Dymetic 1223K.

Structurally flow sensor consists of four acoustic transducers, arranged by two diametrically pairs in section perpendicular to the pipeline axis.

The principle of operation flowmeter is based on measuring the transit time of the perturbed "element" of the stream (heterogeneity caused by turbulent fluctuations) between two pairs of the transducers with the known distance between them. During operation flowmeter, two acoustic emitters (Emitter 1 and Emitter 2) excited by ultrasonic frequency generators emit ultrasonic vibrations, which, after passing through the gas flow, are perceived and converted into electrical signals in the secondary acoustic receivers (Receiver 1 and Receiver 2).

Due to the interaction of ultrasonic beams with inhomogeneous in the gas flows secondary oscillations are modulated in phase. The modulation signal is allocated by the phase detector. Then time of the heterogeneity between sections 1 and 2 is defined. The calculator of flowmeter with known times and a given geometric dimensions (inner diameter of the pipeline, the distance between sections 1 and 2, the roughness of the pipeline walls) calculates the average speed, instantaneous flow and generates normalized frequency output signal proportional to the volumetric flow.

A positive feature of this method is its relative simplicity and reliability of the physical method and the absence of the influence of the gas composition on the determination result of the flow velocity. The disadvantages - a narrow dynamic range.

The principle of the thermal-mass flowmeter TurboFlow TFG-S is that the medium washing the heated temperature sensor, carries away with itself certain part of the thermal energy transducer. In the thermal flowmeters medium bends around two temperature sensors. One sensor provides the reference value - the temperature of the medium. The second sensor is a heated element that gets exactly as much energy to compensate for the thermal dissipation and maintain a well-defined temperature difference. The larger the mass flow washes heated sensor, the more heat energy is dissipated, and the more power must be supplied to the sensor to maintain a temperature difference. Thus, the current in the heating circuit of the sensor is measured parameter determining the gas mass flow rate.

Figure 2. Exterior view thermal-mass flowmeter TurboFlow TFG-S.

3. The experiment scheme

Objective: To determine actual flow rate of the APG and the choice of means for measuring its accounting.

At the facility "Prikamneft" "Elabuga UPS" were consistently set measuring computer complex (MCC) APG (reg.№47248-11 in the State Register of Measurement) consisting from ultrasonic flowmeter Dymetic-1223K (correlation method), calculator VTD-U, temperature sensor and absolute pressure (as a control) and the flowmeter TurboFlow TFG-S (as working (test)). Mounting diagram is shown in Figure 3. Collection of information about the flow rate of the APG was conducted for 36 days.

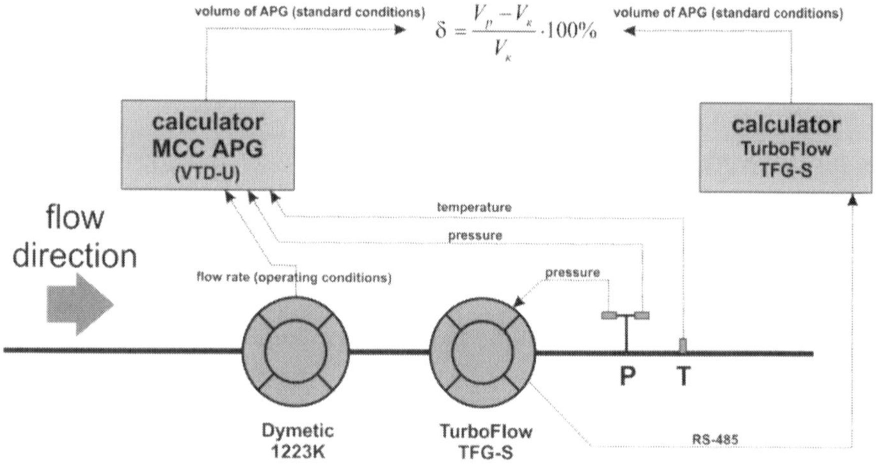

Figure 3. Mounting diagram of measurement instruments

Chosen mounting circuit (Figure 3 and Figure 4 – gas flow direction – from left to right) of measuring instruments excludes the mutual influence (Ultrasonic flowmeter and the pressure sensor do not disturb the flow velocity profile of gas before TurboFlow TFG-S).

Figure 4a. Measuring section with MI flow, pressure and temperature.

Figure 4b. Cupboard power supplies of sensors, data processing equipment, recording and telemetry.

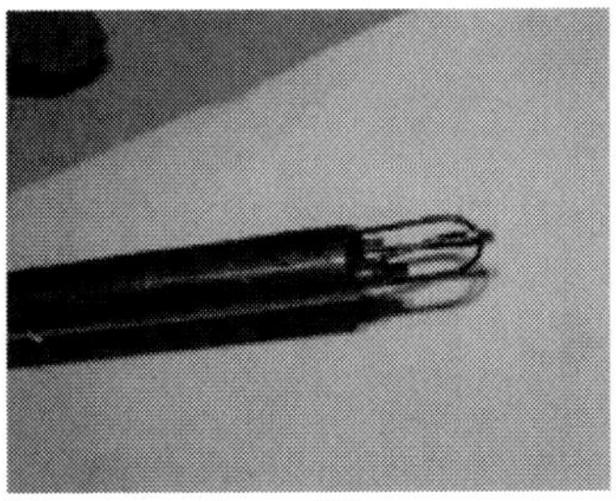

Figure 4c. The sensing element TurboFlow TFG-S.

4. The test results

Below are the diurnal measuring computer complex APG and TurboFlow TFG-S over the observation period (36 days), and error δ *Vp* relative to the volume APG, measured and reduced to standard conditions measuring computer complex APG V_K, calculated by the formula [2]:

$$\delta = \frac{V_p - V_\kappa}{V_\kappa} \cdot 100\%, \tag{1}$$

where *Vp* – volume APG, measured and reduced to standard conditions by flowmeter TurboFlow TFG-S, V_K - volume APG, measured and reduced to standard conditions by measuring computer complex APG.

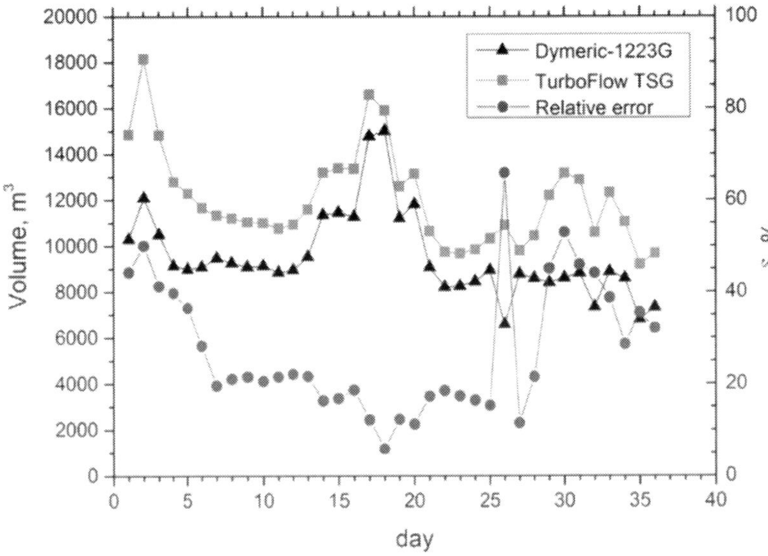

Figure 5. Graph changes daily flow rate and the relative error.

As seen from the graph changes daily flow rate over the period of observation on both devices takes place constant component of the error in determining the flow rate by flowmeter TurboFlow TFG-S, which is caused by the exercise calibration on natural dry gas. And also is present a recurring error caused by the pollution of sensitive elements of the device, which leads to unpredictable disruption dynamics of heat transfer and distortion of results.

5. Conclusions

In the analysis of the obtained results, we can draw the following conclusions: physical method of flow measurement TurboFlow TFG-S is unreliable when used on associated (free) petroleum gas, which is usually is present the second phase in the form of oil vapor, hydrocarbons and water. However, it should be noted that the use of self-cleaning function of the sensing elements of the probe can maintain working capacity of the flowmeter for a long time without intervention of staff.

Thus, when using a thermal-mass flowmeter for the wet free gas it is necessary to consider that the heat and mass transfer characteristics vary significantly depending on the composition and properties of the medium. Therefore, need to calibrate the flowmeter at the same gas composition and properties, or the development and application of a mathematical algorithm in determining the speed (flow rate) of the medium at its various compositions and properties.

The work is performed according to the Russian Government Program of Competitive Growth of Kazan Federal University.

References

[1] GOST R 8.733-2011 *System parameters and measurements of the amount of free gas. General metrological and technical requirements* (Moscow: Standartinform) p 32

[2] GOST R 8.615-2005 *Measuring the amount extracted from subsurface oil and gas. General metrological and technical requirements* (Moscow: Standartinform) p 23

IOP Conf. Series: Materials Science and Engineering **69** (2014) 012034 doi:10.1088/1757-899X/69/1/012034

Creation of anisotropic friction conditions during the process of hot forging

D L Pankratov, R I Valiev
Kazan Federal University,
423800, 68/19 Mira avenue, Naberezhnye Chelny, Russia

E-mail: chelny@kpfu.ru

Abstract. The features of changing the form of ring samples during the upsetting depending on the billet and the tool surface friction conditions are considered.

1. Introduction

In all the laws, describing the boundary conditions during the process of plastic deformation in one form or another, there is a concept of friction coefficient, which is defined only by isotropic friction conditions on the surface of contact of the stamp and a deformable workpiece [1; 2; 3; 4]. However, in real conditions, it is necessary to distinguish roughness along the machining marks - longitudinal and across them - transverse [5; 6]. That is why the coefficient of friction, which is largely dependent on the surface roughness of the tool during the plastic deformation in two mutually perpendicular directions, has different values. Therefore, it would be more precisely to use two friction coefficients in the description of boundary conditions - parallel and perpendicular to the machining marks of the tool surface. In this regard, the shape of the workpiece after deformation may vary depending on the values of the friction coefficient and anisotropy [7].

2. Basic part

This is especially relevant during the process of upsetting of the ring billets.

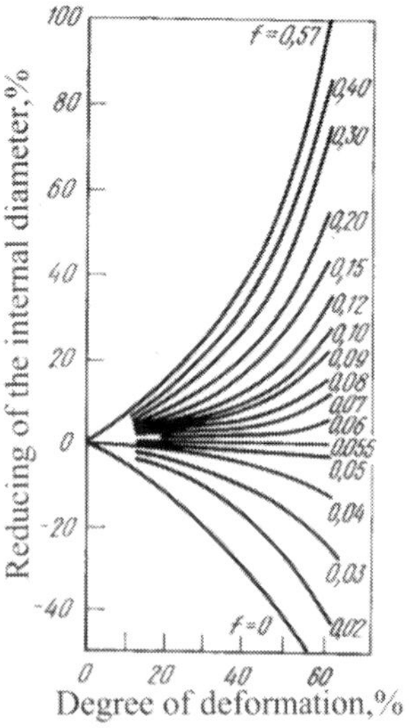

Figure 1 - Curves for determining the friction coefficient using the method of ring upsetting [8; 9; 10, 11, 12].

Based on the analysis of methodology for determining the friction coefficient (Figure 1) proposed by Cockcroft MG, it is possible to distinguish the following options of form changes of the ring samples during the process of upsetting depending on the obtained coefficient of anisotropy *I*:When I = 1 (isotropic friction) and the friction coefficient is equal to 0 the flow of metal goes to the radial direction from the axis of symmetry of the ring (Fig. 2a). Such character of form changing is observed in the range of $0 < \mu < 0{,}055$. In this case, the area of flow division in contrast to the ideal state, when the area of flow division was at the point, represents a circle which diameter is smaller than the internal diameter of the ring sample (Fig. 2b). When $\mu = 0{,}055$, the area of flow division reaches the internal diameter of the sample that leads to the increase only of the external diameter of the ring at the constant internal diameter (Fig. 2c). Increasing $\mu > 0{,}055$ leads to the movement of the area of flow division inside the ring sample. In this case, two types of plug flows are observed (Q1 - of the center and Q2 – towards the center of the ring sample) (Fig. 2d). To get the flow of metal Q2 during the upsetting of the ring samples to the axis of symmetry of the ring, it is necessary to provide the imposition of external restrictions on the side surface of the ring sample when there are large μ values (Fig. 2e).

Section 1.01

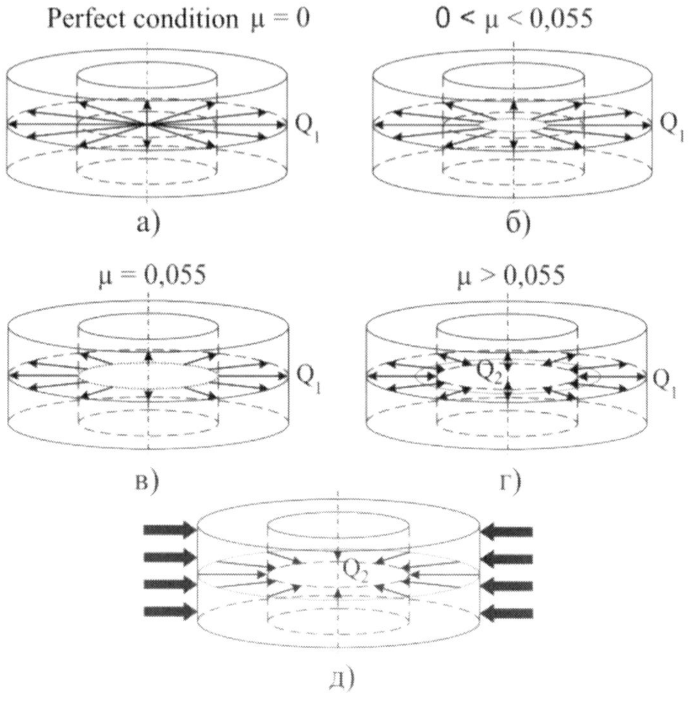

Figure 2 - Schemes of the egress of metal during the process of upsetting of the ring samples for different values of contact friction

When I < 1 (anisotropic friction), the appearance of the ring samples in the form of ellipsoid during the upsetting can be observed. There are four possible variants differ from each other by the

IOP Conf. Series: Materials Science and Engineering **69** (2014) 012034 doi:10.1088/1757-899X/69/1/012034

scheme of the flow of metal. In the case, when the friction coefficients along and across the machining marks are more than $0 < \mu_{//} < \mu_{\perp} < 0{,}055$ the form change is observed, when the flow of metal move of the center of the sample, which leads to the increase in both external and internal diameter of the ring. As friction forces act on the contact surfaces, being slightly different in absolute value, it results in a small ellipse of the external and internal surfaces of the sample in the direction of the machining marks of the tool surface (Fig. 3a). The area of the flow division in this case is less than the internal diameter of the ring, but in contrast to the isotropic conditions it takes the form of the ellipsoid character.

When the friction coefficients along and across the machining marks are more than $0 < \mu_{//} < \mu_{\perp} = 0{,}055$ the form change is observed, at which the flow of metal move of the center of the sample, like in the first case. However, the internal diameter of the ring pattern in the direction that is perpendicular to the machining marks remains constant and increases along the machining marks. The external diameter of the ring increases both along and across the machining marks.

Herewith, the change in diameter along the machining marks is more significant than across. As a result, the ellipsoid character of both internal and external surfaces of the upsetting ring in the direction of the machining marks is increasing (Fig. 3b) as compared with the case in Figure 3, a.

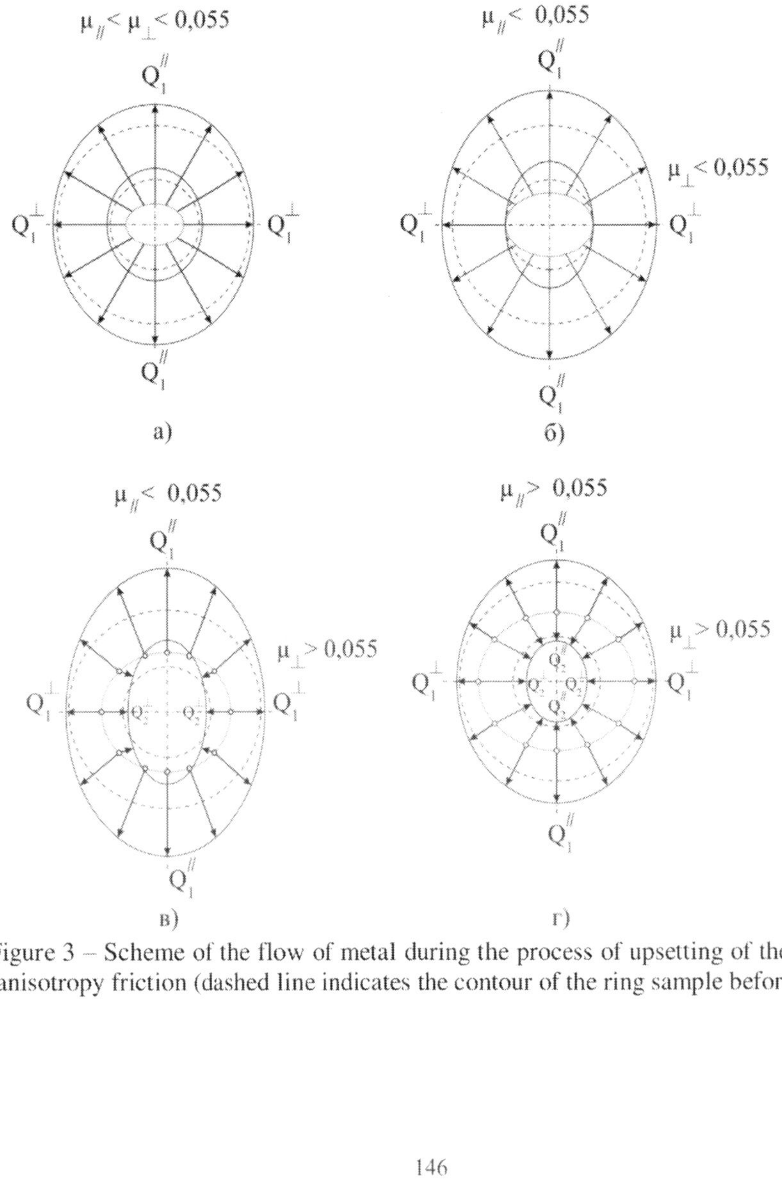

Figure 3 – Scheme of the flow of metal during the process of upsetting of the ring samples in terms of anisotropy friction (dashed line indicates the contour of the ring sample before deformation).

When the friction coefficient along the machining marks $\mu_{//} < 0{,}055$, and across the traces of processing $\mu_{\perp} > 0{,}055$, the area of the flow division is partially within the sample, and partly in the hole of the ring (Fig. 3c). This leads to the appearance of oppositely directed plug flows of metal towards the external ($Q^{\perp}_{1}$) and internal ($Q^{\perp}_{2}$) surface of the sample in the direction that is perpendicular to the machining marks of the tool contact surface. Displaced flow ($Q^{//}_{1}$) can be observed along the machining marks of the tool surface. This leads to a further increase of ellipse of both internal and external surfaces of the ring, as compared with the case shown in Fig. 3 b.

In the fourth case, when a friction coefficient along and across the machining marks is described as $\mu_{\perp} > \mu_{//} > 0{,}055$, the surface of the flow division is within the sample, and its section has the form of ellipse (Fig. 3d). Differently directed plug flows take place throughout the ring sample of the surface of the flow division towards the external and internal surfaces of the ring. This leads to the reduction of internal and external diameters of the sample. The magnitude of ellipse in this case depends on the ratio of the friction coefficients along and across the machining marks of the tool surface, but in any case, this magnitude is smaller than for the scheme in Fig. 3, c.

Suggested hypothesis about the flow of metal during the process of upsetting of the ring samples are in good agreement with the experimental results and allow by means of application of the oriented roughness on engraving of the stamp to get the desired direction of the displacement of metal during the upsetting. The obtained results also can be used to restore worn parts with plastic deformation using the operation of upsetting, stamp dressing, in the case of the defect in the form of insufficient feeling of the stream during the pressure forming. Therefore, to create the zones of hindered flow, it is necessary to direct the machining marks on the engraving of the tool perpendicular to the flow of metal during the plastic deformation. Deformation should occur without the application of technological greasing. The appearance of the slip zones is possible if the machining marks on the engraving of the stamp are directed parallel to the flow of metal during the plastic deformation. The process should be accompanied by the use of technological greasing.

3. References

[1] Storojev M V, Popov E A 1977 *Theory of metal forming*: Textbook for universities. - 4th ed., rev. and add. - M.: Mechanical Engineering 423 p

[2] Kolmogorov V L 1986 *Mechanics of metal forming*: Textbook for universities. - Moscow: Metallurgy 688 p

[3] *Contact friction in metal forming processes* Levanov A N, Kolmogorov V L, S P Burkin, B R Kartak, Ashpur Y V, Spassky Y I – Moscow: "Metallurgy", 1976, 416 p

[4] Severdenko V P, *Theory of metal forming*. - Minsk: High school, 1966. – 233 p

[5] P Poloukhine, V Tiourine, P Davidkov, D Vitanov. 1987 *Troitement des metaux par deformation*. – Mir 318 p

[6] Matalin A A 1985 *Engineering Technology*: Textbook for engineering universities for specialty "Technology of mechanical engineering, machine tools and instruments." - Mechanical engineering, Leningrad Branch 496 p

[7] Schuster L S 1994 *Basics of triboengineering*. – Ufa: Ufa State Aviation Technical University 107 p

[8] Cockroft M G. Greases and lubricating materials: Greases in metal forming processes: Trans. from English. - Moscow: Metallurgy, 1970. - 111 p.

[9] Male A T, Cocroft M G 1966 *Journal of the Institute of Metals* vol. 93 P 38-46

[10] Grudev A P and others. 1982 *Friction and lubrication in metal forming*: Moscow: Metallurgy 312 p

[11] Abdullin I Sh, Kashapov N F and Kudinov V V 2001 Fizika i Khimiya Obrabotki Materialov **6** 69-75

[12] Abdullin I Sh, Kashapov N F and Kudinov V V 2000 Fizika i Khimiya Obrabotki Materialov **3** 35-38.

Innovative Mechanical Engineering Technologies, Equipment and Materials-2013 IOP Publishing
IOP Conf. Series: Materials Science and Engineering **69** (2014) 012035 doi:10.1088/1757-899X/69/1/012035

Imrovement of the cold forming technology of the parts such as longeron

L R Kashapova, D L Pankratov. A A Bilyalova
Kazan Federal University,
423800, 68/19 Mira avenue, Naberezhnye Chelny, Russia

E-mail: chelny@kpfu.ru

Abstract. As a result of modeling in LS-PREPOST preprocessor of the program LS-DYNA a range of radii of curvature edge transition matrix (27,5 ° ≤ R ≤ 48 °) is obtained, which allows to produce defect-free stamping slots for longeron shock absorbers of tractors KAMAZ-5460.

Introduction

Production of high-quality parts for today is one of the main tasks of the leading engineering enterprises. The particular interest represents the frames with curved in a vertical plane longerons, constituting the cargo carrying part of the vehicle. These spars are arching in the locations of the front and rear wheels, which provides large stroke wheels, lowers the center of gravity of the car and, therefore, increases its stability. Spars, curved in the vertical plane, used in the construction of tractors such as, for example, models KAMAZ 5460

Method

The complexity of manufacturing of the parts is in the occurrence of defects in the recesses places: in the tension zone (there is a rupture of the sheet and dent) and in the zone of coagulation (excessive tightening of metal). These zones are shown in Figure 1.

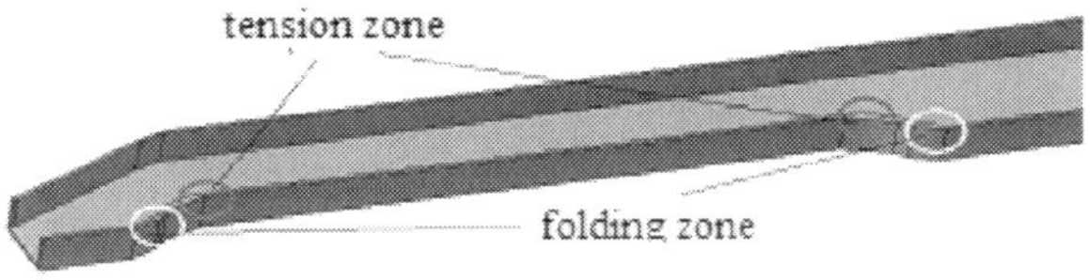

Fig. 1. Area defects on the spar

The reasons of the defects in cold forming can be: defects of the raw material, discrepancy between the material mechanical properties and technical requirements, thickness, formability, surface quality, insufficient detail manufacturability; imperfect design stamps or improper exploitation; improperly developed technology, etc. [1]

Improperly designed blanks and semi-finished products in the die, improper form and sizes of the punches, matrixes, feeding, fixing and pressing parts, insufficient or excessive clamping force, small rigidity of the stamp – all this may be the cause of such types of defects, as variation in wall thickness, folds, teaser, anguishes, unfinished stamping, material gaps, etc.[2]

The sheet material to be stamped, usually has anisotropy of mechanical properties due to trademark of material and technological modes of its receipt, which can have both positive and negative effects on the course of sustainable technology of metal forming processes under different temperature and speed regimes of deformation. [3, 4]

In the process of forming by pressure treatment, as the die cavity filling, in some areas of a blank the deformation path varies. [5]

Causes of defects can also be incorrect or inaccurate installation stamp wear and damage of its parts, weakening and improper adjustment springs. [2]

So-called recess under shock absorbers obtained by the sheet metal forming operation, 8 mm thick. The depth of recess is 70 mm.

In the production the raw material is subjected to incoming inspection and spar is tech part (its shape is simple enough), therefore a possible cause of the above defects is imperfect design stamps, namely irrational geometry of deformation element stamp.

Rational choice of tool geometry will ensure production of high quality parts.

The spar is obtained by the forming process, however, the mechanism of deformation in this case can also be viewed as single point extraction operation. The most critical parts of the working stamp elements are their surfaces wetted during drawing with stamped the material: rounding, transitional chamfer, thresholds, etc.

Metal forming processes today are among the most advanced technologies in engineering all over the world, because provide high quality products at low resource consumption. Metal forming processes technology has high rates of material use. [6]

The use of various computer programs for modeling processes allows to improve metal forming processes technology.

For calculating processes it is often used modeling of technological processes applying finite element modeling software. The programs, which are used in the calculation of production processes in metal forming are the LS-DYNA, ANSYS, ABAQUS, etc. [7]

Purpose of the calculations is to improve manufacturing spar truck KAMAZ-5460 techniques with the help of molding operation computer simulation. Modeling was performed using the software package LS-DYNA and preprocessor LS-PREPOST, in the result the influence of parameters of the working parts of the instrument for a metal (corner radius edge transition matrix R) was investigated.

LS-DYNA - a multipurpose program that uses the explicit wording of the finite element method (explicit finite element method) - is designed to analyze the nonlinear dynamic response of three-dimensional inelastic structures. [8] The Finite Element Method - numerical method for solving complex engineering problems. [9] It was originally used in studies aimed to research and study of electromagnetic fields. [10]

Grid of the blank and instruments surfaces finite element is created in the licensing program UGS NX 7.5 and imported into the preprocessor LS-PREPOST of LS-DYNA program. It is presented in Figure 2.

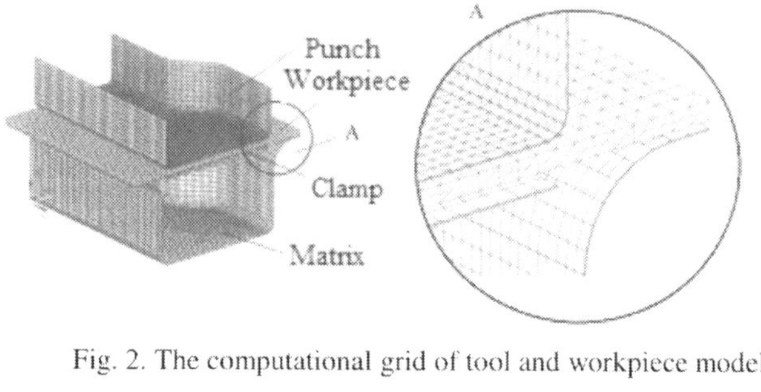

Fig. 2. The computational grid of tool and workpiece models

Tool on the core technology and thinning rate while simulating in LS-DYNA program are shown in Table 1.

When drawing the details of the material with the thickness exceeding 6 mm radius of curvature of the matrix must meet the following conditions:

$$R = (2...4) \times S,$$

where S – material thickness, mm [11].

In this case, S = 8 mm. Therefore, the range of radii of curvature of the matrix will be:

$$R = (2...4) \times 8 = (16...32)\text{мм}.$$

We represent the obtained condition in the form:

$$16 \le R \le 32\text{мм}.$$

In some cases, when S> 3 mm, this condition is as follows:

$$R = (4...6) \times S,$$

where S – material thickness, mm [11].

Then the range of radii of curvature of the matrix will be:

$$R = (4...6) \times 8 = (32...48)\text{мм}.$$

We represent the the condition obtained in the form:

$$32 \le R \le 48\text{мм}.$$

Then we obtain a range of radii of curvature of the matrix in this case:

$$16 \le R \le 48\text{мм}.$$

Considering the obtained condition for the following calculations are selected transition edge radii of Rcurv matrix.: 20, 25, 30, 35, 40 and 45 mm.

We should keep in mind that the size of the clip A and B have been changed, due to insufficient clearance between the workpiece and the matrix. Previews of the clip and the matrix are presented in Figures 3 and 4, respectively.

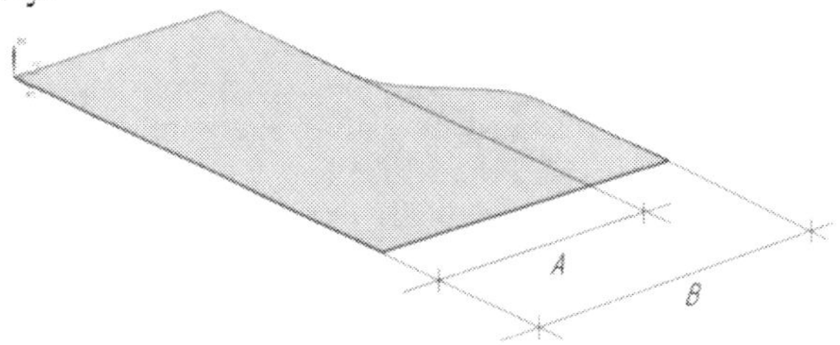

Fig. 3. Drawing of the clip with A and B size designation

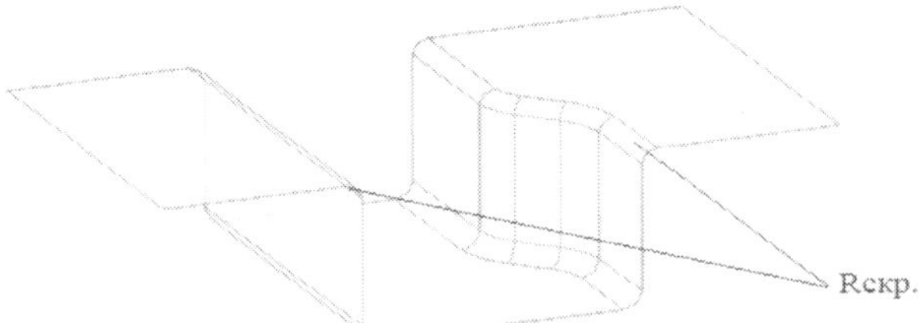

Fig. 4. Drawing of the matrix with the specified radius of curvature $R_{скр.}$

In the result of the calculations in the program LS-DYNA the following thinning values φ were obtained:

а) a) following the underlying technology (A = 191,2 mm, B = 261,2 mm, Rcurv. = 18mm) φ=64%;

б) b) following the design technologies (A = 181,2 mm, B = 251,2 mm Rcurv. = [20, 25, 30, 35, 40 and 45] mm) φ=[58; 88; 20; 18; 19 и 18] % accordingly.

To determine the relationship between the flow and shape of the deforming metal tool is needed to construct curves in the Cartesian coordinate system using the data obtained from calculations.

We represent the a thinning change from simulation parameters as functions:

$L_R = f(R)$ – distribution thinning values curve φ depending on the radius of edge curvature of matrix R.

$L_{1\varphi]} = 35\%$ – curve of allowable thinning values *φ for steel* ([φ]=35°).

$L_1 = 16$, $L_2 = 48$ – curves expressing the condition of limiting the radius of curvature of the matrix edge.

Curves are shown in Figure 5.

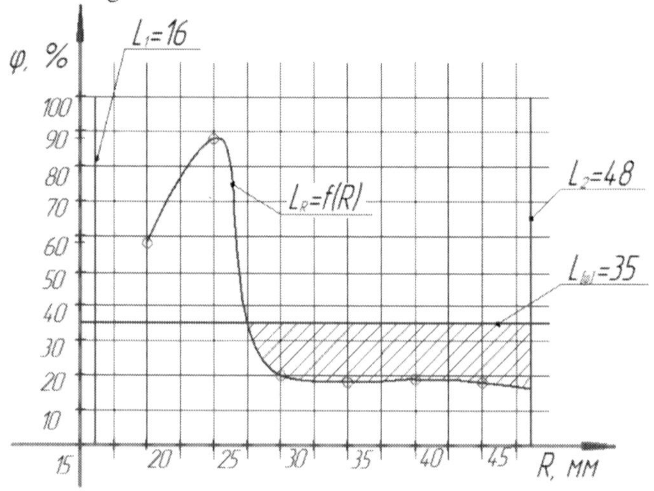

According to the obtained graphs we define the area of rational parameters of tools for defect-free sheet metal stamping by forming operation. For options when changing the radius of curvature of the matrix, the range of radii will be as follows: 27,5°≤R≤48°.

Calculation of the processes using the software allows you to explore the influence of parameters on the working parts of tools for metal and to create optimal conditions for high-quality parts from sheet material.

References

[1] Shukhov Y 1977 *Cold stamping: Textbook* - Moscow: Higher School 208 p

[2] Kilov A S 2004 *Manufacturing of preforms.* Stamping: Series of six textbooks. Book 2. Producing blanks from the sheet and roll-formed profiles - Orenburg: SEI OSU. – 182 p

[3] Golenkov V A, Dmitriev A M, V D Kuhar and others 2004 *Special processes and forming equipment* M.: Mechanical Engineering 464 p

[4] Galyautdinov R T and Kashapov N F 2001 Iz. VUZ. Aviatsionnaya Tekhnika **2** 80

[5] Byeloborodov S G 2010 *Modeling of metal forming processes* / / Materials of the International Scientific and Technical Conference AAI "Automobile and Tractor construction in Russia: Development Priorities and training", dedicated to the 145 anniversary MSTU "MAMI". - M.: MSTU "MAMI" p 47-52

[6] http://www.dynaomd.ru/

[7] http://www.cae-services.ru/

[8] Calpin Y G, V A Ryabov, Anyukhin A S 2010 *Resistance of metal deformation in the complex trajectory of loading* / / Materials of the International Scientific and Technical Conference AAI "Automobile and Tractor in Russia: Development Priorities and training", dedicated to the 145th anniversary of Moscow State Technical University "MAMI". - M.: MSTU "MAMI" p 139-140

IOP Conf. Series: Materials Science and Engineering **69** (2014) 012035 doi:10.1088/1757-899X/69/1/012035

[9] Singiresu S Rao 2004 *The Finite Element Method in Engineering*. Int: Butterworth Heinemann pp 663
[10] Allen Glisson 1994 *The Finite Element Method in Electromagnetics* // IEEE Antennas and Propagation Magazine, Vol. 36, No. 3, June
[11] Rudman L I, V L Marchenko, A I Zaychuk etc. 1988 *Reference of stamps designer.* Stamping M.: Mechanical Engineering 496 p.

Application of the device of slices for automation of matrix calculations

E A Romanovsky
Kazan Federal University,
423800, 68/19 Mira avenue, Naberezhnye Chelny, Russia

E-mail: chelny@kpfu.ru

Abstract. The basic concepts of the device of slices for the processing of multidimensional numerical arrays are represented. The data on the developed matrix component supporting the concept of simple and generalized slices are given. The application efficiency of the slicing device and matrix components in industrial automation at the stages of matrix computation automation is shown.

1. Introduction

Application of the slices device is primarily based on the use of one-dimensional arrays for storing the multidimensional numerical data. The element of these multidimensional numeric data can be as an ordinary real number (double type in the language C + +), as a rather complicated mathematical object of a certain class, for example, a class of super-long rational numbers or quaternion class [1].
Instrumentation of slices showed its high efficiency even in the FORTRAN language, as a key concept for vectors. Some sort of the slices device is implemented in such a well-known computer simulation environment as MATLAB, where, as it is known, multi-dimensional numeric arrays are also stored in memory as one-dimensional arrays [2; 3; 4; 5]. Because of its importance the base classes of the slices device were included into the latest standard of the C + + language in the framework of a set of basic software tools of vector arithmetic [6]

2.Basic part

Application of the slices device for information processing can significantly reduce the number of operations for the dereference of pointers and de-addressing of numeric data. Also, the use of slices can significantly extend the capabilities of parallel and batch processing of numerical data. However, in the field of industrial automation in the processing of information, slices, unfortunately, have not found wide acceptance.
Slices can be divided into ordinary (simple) and generalized.
An ordinary slice is an abstraction that allows you to effectively manipulate one-dimensional array-vector as an array of arbitrary dimensions and sizes, such as matrix.
An ordinary slice always has to do with some vector-array and is described by a triple of integers:
1. The starting index of the vector.
2. The length of the slice.
3. The pitch, which is the number of vector elements between two elements of the slice.
If the starting index of some one-dimensional array-vector for some slice is 2, the length of the slice is equal to 5 and the pitch is 3, then this slice can be described by the expression: slice (2, 5, 3). If we add to the number 2 (starting index) the number 3 (pitch), we get the number 5. If we add to the resulting number 5 again the number 3 (pitch), you get the number 8. If you do it 5-1 times (where 5 is the length of slice), you get a set of indices of the vector-array: 2, 5, 8, 11, 14.
As can be seen from the given example, an ordinary slice always describes a sequence of integer indices and is mapping a set of three non-negative integers into the indices of the array-vector.
Let there be given a matrix with dimensions 4 x 3, the graphical representation of which is shown in figure 1.

00	01	02
10	11	12
20	21	22
30	31	32

Gig. 1. Graphical representation of the matrix

In the language of C ++, multi-dimensional array (in particular, the matrix as well) is convenient in terms of use and efficient in terms of speed to store in memory as one-dimensional array with the storage for additional variable sizes of this multi-dimensional array for each of its dimension [6; 7; 8]. Representation of the reduced matrix in the form of one-dimensional array is possible by the sequence of the elements of the matrix by columns or by sequential arrangement of the elements of the matrix by rows. In the case of a sequential arrangement of the matrix elements by rows an array-vector is obtained, the graphic representation of which is shown in picture 2.

00	01	02	10	11	12	20	21	22	30	31	32
0	1	2	3			6			9		

Fig. 2. Graphical representation of the array-vector

Let it in the form of such an array-vector matrix, shown in fig. 1, be stored in the RAM.

It is not difficult to see that the line shown in fig. 1 of the matrix with the number x can be described using the expression $\text{slice}(3\cdot x, 3, 1)$. Here, the first element of a row x is the $(3\cdot x)$ element of the array-vector, the second element of the row x is the $(3\cdot x+1)$ element of the array-vector and the third, the last element of the row x is the $(3\cdot x+2)$ element of the array-vector. Thus, the expression $\text{slice}(3\cdot 0, 3, 1)$ describes the indices of the matrix row $(00, 01, 02)$, and the expression $\text{slice}(3\cdot 2, 3, 1)$ describes the indices of the matrix row $(20, 21, 22)$.

Also it is not hard to see that the column of matrix shown in fig. 1 with the number y can be described using the expression $\text{slice}(y, 4, 3)$. Here, the first element of the column y is the y element of the array-vector, the second element of the y column is the $(y+3)$ element of the array-vector, the third element of the y column is the $(y+6)$ element of the array-vector and the fourth, the last element of the y column is the $(y+9)$ element of the array-vector. Thus, the expression $\text{slice}(2, 4, 3)$ describes the indices of the matrix column.

As a result, with two slices it is possible to apply the one-dimensional array-vector, as a two-dimensional, i.e., as a matrix. You can also separately apply to any row or column of the matrix. Obviously, the use of n ordinary slices allows you to apply one-dimensional array-vector, as the n-dimensional array.

Ordinary slices can also be used for the expression of diverse subsets of one-dimensional array-vector. So, you can use the ordinary slices to manipulate contiguous sub array within a one-dimensional array-vector. However, the capabilities of ordinary slices are limited. Restriction appears, for example, when it is necessary to select in the matrix a sub array that is not a single row or column of the matrix. For example, you may need to select the sub-matrix with the size 3×2 of the upper-left corner of the matrix with the size 4×3, as shown in fig. 3.

00	01	02
10	11	12
20	21	22
30	31	32

Fig. 3. Graphical representation of the matrix with a selected sub-matrix

Unfortunately, the elements of the sub-matrix are arranged so that they cannot be selected using an ordinary slice (see fig. 4).

00	01	02	10	11	12	20	21	22	30	31	32
0	1		3	4		6	7				

Fig. 4. Graphical representation of the array-vector

To solve this problem you should use the generalized slices.

A generalized slice contains almost all the information from n ordinary slices. Applying the generalized slice containing two pairs of non-negative integers (length, pitch), we can select two-dimensional sub array (sub matrix) within the two-dimensional array (matrix). Applying the generalized slice with three pairs of non-negative integers (length, pitch), we can select three-dimensional sub array within the three-dimensional array, etc.

In addition, each slice contains one generalized non-negative number, determining the initial index of the vector. This index defines the beginning of the sub array within the array.

A generalized slice for selecting the sub matrix within the matrix, as shown in fig.3, can be determined using the following starting index and the two pairs (length, pitch):

1. The starting index of the vector - 0.
2. The first pair (3, 3) - allows you to select 3 elements of the column.
3. The second pair (2, 1) - allows you to select 2 elements of the row.

Together, these data allow selecting a sub array with the sizes 3×2.

The same slice can be determined a little differently:

1. The starting index of the vector - 0.
2. The first pair (2, 1) - allows you to select 2 elements of the row.
3. The second pair (3, 3) - allows you to select 3 elements of the column.

The shown permutation of pairs affects only the order of application of the matrix elements. So, in the first case, application of the matrix elements allows to select the sequence (00, 01, 10, 11, 20, 21). In the second case, application of the matrix elements allows to select the sequence (00, 10, 20, 01, 11, 21).

Thus, the device of slices allows you to select a variety of multidimensional sub arrays within multidimensional arrays (in particular, the sub matrix within matrices) for the effective implementation of many different mathematical operations.

As a programming environment for creating a matrix component, supporting the device of slices, was chosen the environment Microsoft Visual C ++. Besides, at the stage of working out the component, the technology COM has been widely used [9]. The support by the matrix component of slices has been implemented using a well-known and well-proven library ATL (Active Template Library) [10], which is based on a wide use of the COM technology.

For the development of the application, testing the created matrix component, the programming environment Borland C ++ Builder has been selected. To demonstrate the process of testing the component the library OpenGL has been used, that allowed to simulate and reproduce on the screen the required three-dimensional scenes.

Fig. 5 shows the results got by using the OpenGL library graphic-ray imaging applying the generalized slice to the three-dimensional sub array of a three-dimensional array.

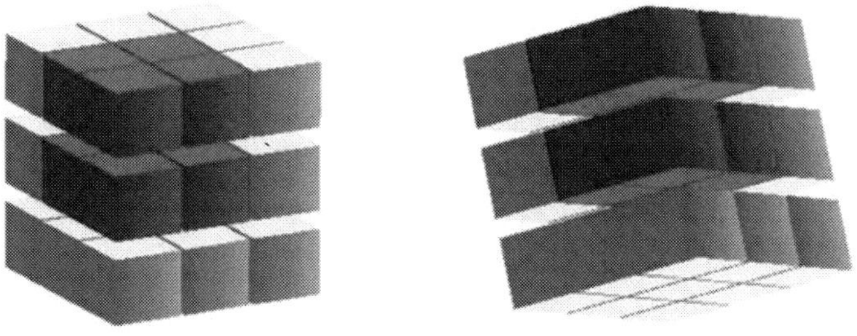

Fig. 5. Visualization of applying the sub arrays using
a generalized slice for a three-dimensional array

The developed component can be used in the process of automating various matrix calculations. If the client application is implemented by mapping a n-dimensional array into a one-dimensional, then using the class of generalized slices, you can implement a n-dimensional slice of the array. So, using the $n = 2$ generalized slice a sub matrix can be selected of a matrix, and you can execute its processing separately from the whole matrix.

The created matrix component has been successfully applied in the software solution of problems of optimal in terms of quadratic criterion of quality of the object management, described by a system of ordinary differential equations. Also, the developed matrix component has been successfully used in the process of working out the tools for numerical solutions of matrix equations of Sylvester, Lyapunov and Riccati.

4.References

[1]. Tan K. Sh., Stib W.-H., Hardy J. *Symbolic C + +: Introduction to computer algebra using the object- oriented programming*. 2001 Trans. from the 2nd Eng. ed. – M.: Mir, 622 p.

[2]. Martynov N N, Ivanov A P. MATLAB 5.x. *Computation, visualization, programming*. 2000 M.: KUDITS - IMAGE , 336 p.

[3]. Potemkin V G *System of engineering and scientific calculations of MATLAB 5.x: In 2 volumes*. Volume 1. 1999 M.: DIALOG- MIFI, 366 p.

[4]. Potemkin V G *System of engineering and scientific calculations of MATLAB 5.x: In 2 volumes*. 1999 Volume 2. - M.: DIALOG- MIFI, 304 p.

[5]. Medvedev V S, Potemkin V G *Control System Toolbox. MATLAB 5 for students* / under total. Ed. Cand.T.S. V.G. Potemkin. M.: DIALOG- MIFI. – 287 p.

[6]. Stroustrup B *Programming Language C + +*, spec. ed. / Transl. from English. 2002 - M.; St. Petersburg.: "Publisher BINOM " - " Nevsky Dialect ", 1099 p.

[7]. Lippman S, Lazhoye J Programming language C + +. *Introductory Course, 3rd ed.* / Transl. from English. 2003 St. Petersburg. - M.: Nevsky Dialect - DMK Press, 1104 p.

[8]. Sedgwick Robert. *Fundamental algorithms in C + +. Analysis / Data Structures / Sorting / Search:* Trans. from English. / Robert Sedgwick. 2002 - St. Petersburg. LLC " DiaSoftYuP ", 688 p.

[9]. Rogerson D *Fundamentals of COM* / Trans . from English. 1997 M: Russian Edition , 228 p.

[10]. Troelsen E *COM Model and use of ATL 3.0* / Trans. from English. 2005 - St. Petersburg.: BHV-Petersburg, 928 p.

IOP Conf. Series: Materials Science and Engineering **69** (2014) 012037 doi:10.1088/1757-899X/69/1/012037

Pulse Plasma Surface Thermostrengthening of Machine Parts

I H Israfilov, A T Galiakbarov, D A Bashmakov, A T Gabdrahmanov, A D Samigullin
Kazan Federal University,
423800, 68/19 Mira avenue, Naberezhnye Chelny, Russia

E-mail: chelny@kpfu.ru

Abstract. The paper presents a pulse plasma generator for surface hardening of parts. The given equations allow to calculate the parameters of the pulse plasma generator to ensure the specified quality indicators of the hardened zone.

1. Introduction

Thermal processing of metals leads to modification of their structure, which affects the operational characteristics of products. In many cases, the local heat treatment is technically and economically justified. Thus reinforce the most loaded working surface, at a constant microhardness of the matrix. The concentrated streams of energy are widely spread for surface hardening of details in industry. Surface heat treatment of the parts with the use of concentrated energy flux is a substantial reserve of saving material and energy costs [1; 2; 3].

Experience shows that the plasmatron as a source of energy for surface heating can be applied in many cases along with such sources as laser and electron-beam, providing more high technical-economic indicators of the technological process[4; 5; 6; 7, 8, 9].

In this paper an effective method of plasma hardening of the surface layer due to the cumulative effects of pulsed-plasma stream is proposed: elastic strain interaction with a shock wave and pulsed plasma jets.

2. Basic part

Pulse plasma generator consists of a water-cooled frame, two electrodes of the same size and shape located in parallel to each other. Plasmatron works the following way: electric arc, moving on the surface of the electrodes in the field of their own current (railotron effect), heats the plasma gas, arc separates from the electrode surface between electrodes, then compressed plasma arc along with the flow of heated gas comes out of the nozzles and gets on the workpiece.

The hardening process is carried out due to rapid heating of the surface layer with the subsequent cooling by heat in the amount of metal as well as in the environment. High speed of heating and cooling of the surface layer of metal (>106 K/s) contributes to the formation of fine crystalline structure with a high dislocation density. By regulating the energy parameters and the duration of the cycles of heating and cooling, as well as the application of various plasma-forming gases, it is possible to change the structural-phase state of the surface layer of metal, creating the optimal structure for required operating characteristics [10].

The main parameter affecting the quality indicators of the hardened area is the temperature zone plasma impact. It should be noted that a reliable way to check the temperature changes in the areas of plasma heating does not currently exist. Therefore, the analysis of thermal processes is analytically on the basis of solutions of the heat equation with the boundary and initial conditions and certain assumptions. In this context, the study and modeling of distribution of temperature in pulse plasma heat treatment is of great importance.

For the determination of temperature fields at pulse plasma action it is accepted that there is no internal heat source in the sample, because of the low energy not only penetration of charged particles doesn't happen, but even of neutral particles into the depths of the material and the thermal wave propagates at a distance of more than 100 microns; pulse plasma flow is axisymmetric, and the radius of heating spot is significantly more than termoline zones; there are no radiation losses from the surface; the temperature of the irradiated material at the initial moment of time and beyond the zone of thermocline equals 0 degree in Celsius.

For calculations we have the equation describing the process of propagation of heat in metal from distributed instant source density Q j/m2, on the adiabatic surface of the half-space [11] (1).

$$T(x,z,\tau)=n\cdot\{\frac{Q\cdot\eta\cdot\sqrt{a}}{2\cdot\lambda\cdot\sqrt{\pi\cdot\tau_{И}}}\cdot\exp\left(-\frac{z}{4\cdot a\cdot\tau_{И}}\right)\cdot\left[erf\left(\frac{x}{\sqrt{4\cdot a\cdot\tau_{И}}}\right)-erf\left(\frac{x-2\cdot r_p}{\sqrt{4\cdot a\cdot\tau_{И}}}\right)\right]+$$
$$+\frac{Q\cdot\eta}{2\cdot\lambda\cdot\sqrt{\pi}}\cdot[\frac{\sqrt{a}}{\sqrt{\tau_{Об}}}\cdot\exp\left(-\frac{z}{4\cdot a\cdot\tau_{Об}}\right)\cdot\left[erf\left(\frac{x}{\sqrt{4\cdot a\cdot\tau_{Об}}}\right)-erf\left(\frac{x-2\cdot r_p}{\sqrt{4\cdot a\cdot\tau_{Об}}}\right)\right]- \quad (1)$$
$$-\frac{\sqrt{a}}{\sqrt{\tau_{П}}}\cdot\exp\left(-\frac{z}{4\cdot a\cdot\tau_{П}}\right)\cdot\left[erf\left(\frac{x}{\sqrt{4\cdot a\cdot\tau_{П}}}\right)-erf\left(\frac{x-2\cdot r_p}{\sqrt{4\cdot a\cdot\tau_{i_П}}}\right)\right]]\}+T_0$$

where n is the number of pulses, Q is the amount of energy per pulse Joule/m2, η-plant efficiency factor, a is the thermal diffusivity coefficient m2/s, λ - thermal conductivity coefficient W/m•K is a function of temperature [12], $\tau_И$ - the duration of the plasma effect on the sample sec, $\tau_П$ - pause time sec, $\tau_П=l_Э/V$, $l_Э$ - length of electrodes m, V - velocity of the arc m/s, $\tau_{Об}$ is the period sec, $\tau_{Об}=\tau_И+\tau_П$, z - depth impact m, x is coordinate $\perp$ z m, r_P – is a spot radius of influence m.

The temperature fields in the sample after exposure to n calculated plasma flux pulses are presented on fig. 1.

The proposed model (1) is a solution of a nonlinear of the first kind which takes into account the change of the thermophysical properties of the material ($\lambda(T)$) depending on the temperature. The model allows to calculate the number of pulses of the given energy and thermal parameters of pulse plasma generator to achieve the required temperature in the zone of influence of plasma with the detail and depth thermostrengthening.

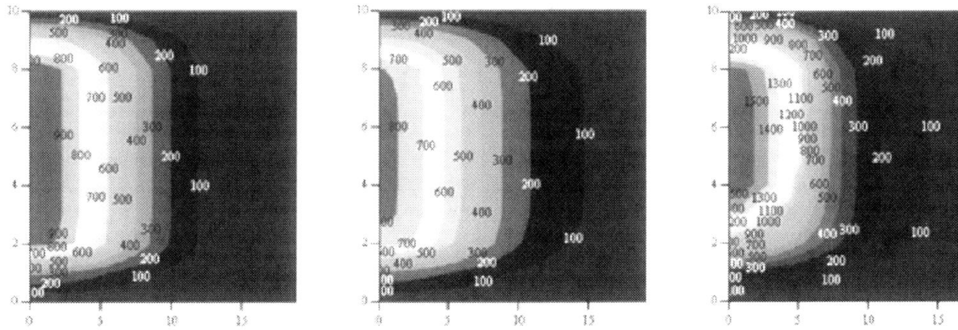

Fig. 1. The calculated temperature field in a sample of steel 45 after 15 pulses; a) without regard to cooling between pulses; b) including cooling between pulses; c) including cooling between pulses and change in λ temperature

To verify the model experimental studies of pulsed plasma flow on the surface of metals were conducted. Various alloy tool steels and steels with different content of carbon were selected as the samples. The purpose of metallographic was a determination of the depth and structure of the hardened zone. Figure 2 presents a sample of steel 45 processed with impulse plasma generator.

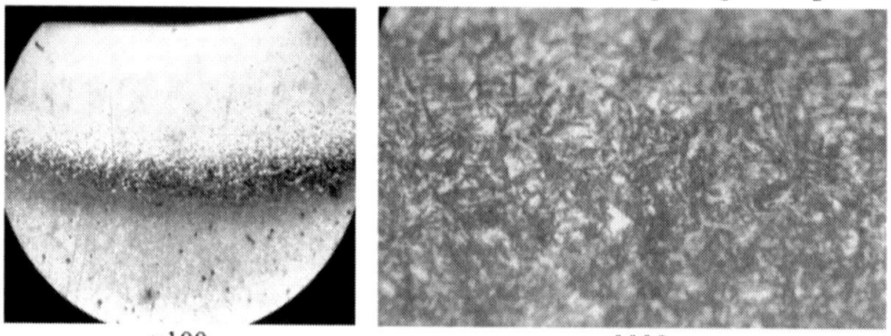

×100 ×1000

Fig. 2. The sample of steel 45 processed with impulse plasma generator, the power P=6 kW, air flow G=124 l/min

Metallographic investigations of different steels in the field of plasma hardening showed that the zone of thermal influence of the plasma jet is the shape of the segment. When metal surface by plasma jet is heated, surface layer is being heated up to different temperatures, as a result it has a layered structure (Fig. 1.) which corresponds to the selected model.

Change of hardness depth of the hardened zone is represented in Fig. 3.

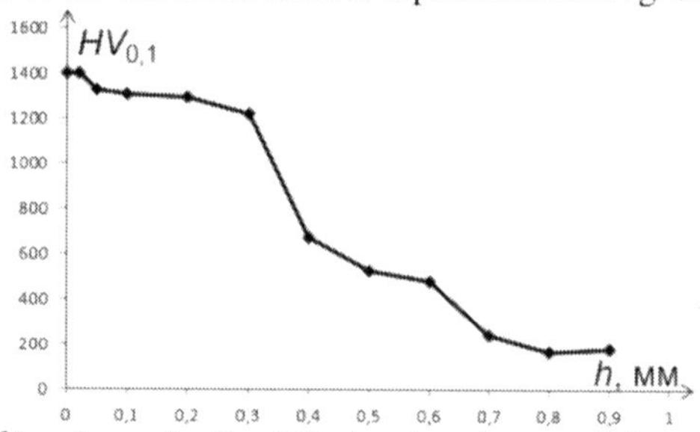

Fig. 3. Change of hardness depth of the hardened zone of the sample of steel 45.

Research has shown that the successful solution of problems of formation of high-strength state of the surface layers of carbon steels is conducting plasma surface hardening. This creates a dislocation structure formed at high temperature thermomechanical treatment and providing high strength and plasticity of the matrix.

The calculations allow to define the necessary parameters of pulse plasma generator to ensure the set of quality indicators of the hardened zone.

3. References

[1] Klebanov Y D, Grigoriev S. *The physical basis for the use of concentrated energy flows in materials processing technologies. / Second edition. Textbook.* 2005 M .: IC MSTU "STANKIN, Janus-to." 220c.

[2] Hasanov I S *The plasma and beam technology* 2007 Baku: Elm - 171 p

[3]Israfilov IH, Rakhimov RR, Khaibullin II, Saubanov RR *Prospective application of highly concentrated energy for surface heat treatment product* 2011 The socio-economic and technical systems: research, design, optimization. T. 58, No. 1 - P. 25-30.

[4] *Encyclopedia of Low Temperature Plasma: Introductory volume to 9 kN. Proc. 4* 2000 Ed. Academician VE Fortova. - M .: Nauka, 516 p.
[5] Shakirov YI, Valiev RI, Hafiz A., GY Shakirov *The multi-channel system with plasma electrolytic cathode* 2011 Automotive 2, - pp 36-38.
[6] *Low-temperature plasma: T.18. High energy processes materials processing*2000 / Exec. Ed. MF Zhukov; Ros. Acad. Sciences, Sib. Dep-tion, Institute of Thermophysics. - Novosibirsk: Nauka, 425 p
[7] Taran V M, Lisowski S M, Lyasnikova A V *Designing elektroplazmennyh technology and automated equipment* 2005 M .: MSTU. NE Bauman, - 250p
[8] Zvezdin V V, Galiakbarov A T, Saubanov R R, Gabdrakhmanov A T, Nugumanova A I *The influence of parameters of pulsed plasma generator on quality Process* 2010Vestn. KSTU. Tupolev. 2 - p. 50-52.
[9] Parkin AA *Processing Technology concentrated energy flows* 2004 Samara: Samara State Technical University, 497 p.
[10] Davydov S V, Gulyaev Y V, Simochkin V V *Effect of thermal properties of carbon steels at eutectoid transformation of austenite* 2008 Bulletin of the Bryansk State Technical University, 1. S. 4-9.

Quantum-chemical calculations of the structure and vibration spectrum of methyl nitrate

R M Shaikhullina, G M Khrapkovskii, F S Sarvarov
Kazan Federal University
423800, Republic of Tatarstan, Naberezhnye Chelny, 68/19 Mira avenue

E-mail: chelny@kpfu.ru

Abstract. Methyl nitrate vibration spectra theoretical analysis data within the B3LYP density functional theory method in the 6 -31G (d) basis are represented. When optimizing the geometry of the molecule it was found out that methyl nitrate is characterized by one stable conformation: *trans* - form. Frequency, intensity and forms of normal vibrations are calculated, their comparative analysis with well-known from the literature experimental data is given.

1. Introduction

Spectroscopy is one of the most informative methods for analyzing the structure and properties of molecules [1; 2; 3]. For the interpretation of the experimental spectra and the analysis of molecular structure theoretical studies based on quantum chemistry methods are widely used in recent years [4; 5; 6; 7].

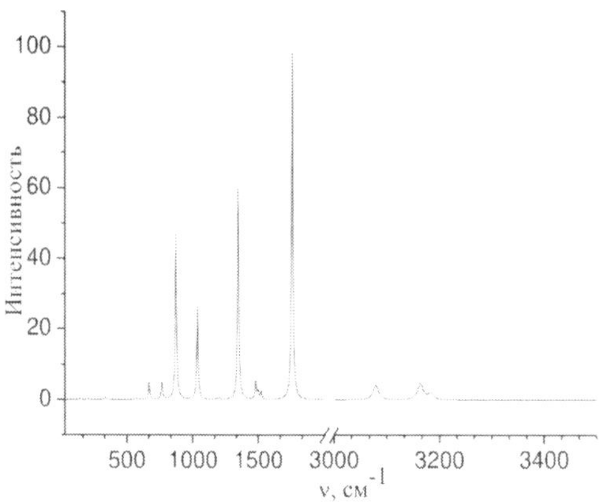

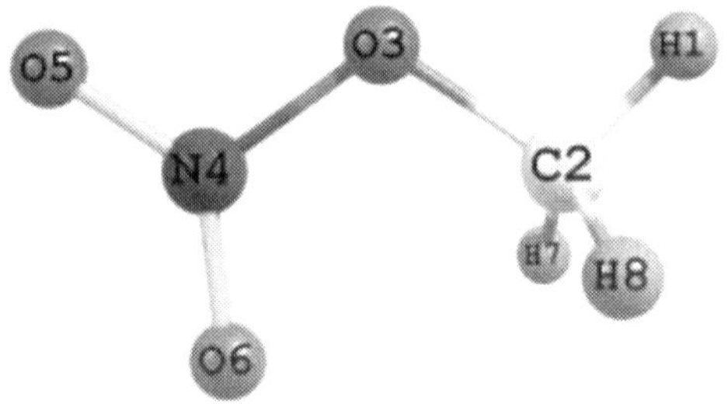

Fig.1. Theoretical IR spectrum and structure of methyl nitrate

In our work we present the data on the theoretical analysis of the structure and vibration IR spectrum of methyl nitrate (MTN) - the simplest representative of a number of aliphatic nitrogen-containing compounds (Fig. 1).

2. Basic part

Quantum chemical calculations were performed using the Gaussian-98 in the framework of density functional theory B3LYP on the basis of 6-31G (d). When optimizing the geometry of the molecules with the scanning of the dihedral angles around the C-O and N-O it was found out that MTN is characterized by one stable conformation: *trans* - form. The atoms of C (2), O (3), N (4) lie in one plane, the atom H (1) is in the *trans*-position relative to the connection N-O, the other two atoms H (7, 8) - in a gauche orientation to the closest oxygen atom. Planar structure of the fragment C-O-NO_2 and the possibility to implement a single conformation is confirmed by the results of earlier studies [8; 9]. The calculated and experimental frequency, intensity and the normal vibrations are given in the table. Spectral characteristics remain unchanged when scanning the dihedral angles, which is another criterion for the absence of additional conformations.

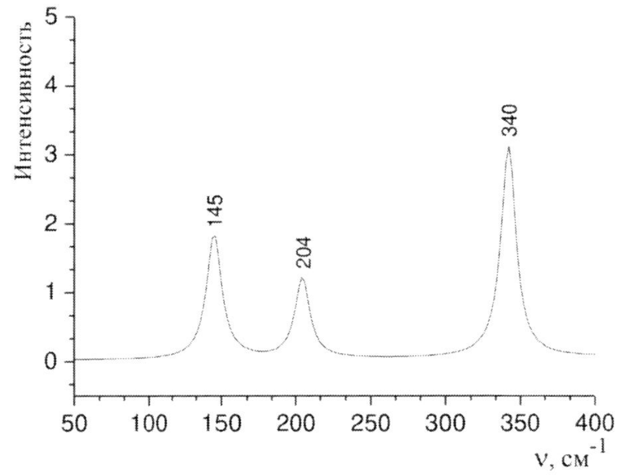

Fig.2. Theoretical methyl nitrate IR spectrum in the frequency range 50 - 400cm^{-1}.

Particular attention is given to the study of low frequency vibrations in the range of 50 - 400cm^{-1}. As it is known, these vibrations are most sensitive to the spatial structure, the molecules pack, intra- and intermolecular interactions. Bending vibrations δ (C-O-N) at 340 cm^{-1} and torsion vibrations τ (C-O) and τ (N-O) at 204 and 145 cm^{-1} are observed in the spectrum of MTN (Fig. 2). The vibration spectra and structure of MTN received by calculations are in agreement with the well-known experimental data and can be used later for the study of more complex nitro compounds, in particular, cellulose nitrates [10].

Table. Experimental and calculated B3LYP/6-31-G (d) frequency and distribution of potential energy of methyl nitrate (using the procedure of the force constants scaling).

Attribution[a]		Calculation		Experiment [8,9]		
The distribution of potential energy ,%		ν, см$^{-1}$	I[б]	Gas	Liquid	Matrix N_2
ν 18	ν_{as} CH_3, ν(C-H1) [76]	3020	0.017	3008	3040	3029
ν 17	ν_{as} CH_3, ν(C-H7)[49], ν(C-H8)[49]	3002	0.046	3008	3040	3029
ν 16	ν_s CH_3, ν(C-H8)[38], ν(C-H7)[38], ν(C-H1) [23]	2921	0.041	2941	2967	2962
ν 15	ν_{as} NO_2 , ν (N=O5) [50], ν (N=O6) [40]	1676	1.000	1678	1634	1656
ν 14	δ_{as} CH_3 - δ(H7CH8)[62], δ(H7CO)[17], δ(H8CO)[16]	1466	0.021	1468	1458	1465
ν 13	δ_{as} CH_3 - δ(H1CH8)[72]	1456	0.022	1455	-	1440
ν 12	δ_s CH_3 - δ(H1CH7)[62], δ(H1CO)[15]	1422	0.054	1430	1429	1434
ν 11	ν_s NO_2 - ν (N=O5) [46], ν (N=O6) [32]	1285	0.598	1287	1282	1291
ν 10	r CH_3 - δ (OCH7)[45], δ (OCH8)[44]	1175	0.005	1176	1174	1177
ν 9	r CH_3 - δ (OCH1)[56]	1158	0.002	1136	1150	1150
ν 8	ν (C-O) [76]	1011	0.267	1017	993	1004
ν 7	ν (N-O), ν (N-O) [43],δ(O=N=O) [32]	862	0.468	884	859	858
ν 6	γ (NO_2) [98]	765	0.049	758	759	763
ν 5	δ(C-O-N) [30], ν (N-O) [29], δ(O-N=O6) [16]	660	0.054	657	662	662
ν 4	r (NO_2) , ν (NO) [51], δ(O-N=O5) [20], ν (CO)[15]	567	0.005	578	577	610
ν 3	δ(C-O-N) [75]	340	0.008	340	-	362
ν 2	τ (C-O) [91]	204	0.003	-	-	-
ν 1	τ (N-O) [96]	145	0.005	-	-	-

[a] ν - bond stretching; δ - bending; r - rocking, γ- not plane, τ - torsion vibrations; as - asymmetric , s - symmetric;
[b] intensities are normalized relative to the maximum.

3.References

[1]. Smith A. 1982 *Applied IR spectroscopy*. Moscow. Mir,

[2]. Zhbankov R G *Infrared spectra of cellulose and its derivatives. Minsk: Science and Technology*, 1983.

[3]. Shlyapochnikov V A *Vibration spectra of aliphatic nitro compounds* 1989 : Nauka 134 p.

[4] Shlyapochnikov V A, Khrapkovskii G M, Shamov A G 2002 *Structure and vibration spectra of mononitroalkanes.* Math. AN Ser.Him. Number 6. Pp.863-870.

[5]. Benjamin J. van der Veken, Maas R, Guirgis G A, Stidham H D, Sheehan T G , Durig J R *Infrared Spectrum, Initio Calculations, Barriers to Internal rotation and Structural Parameters for Methyl Nitrite*. 1990 Vol.94. Pp.4029- 4039.

[6]. Joao Bosco P Da Silva, Nivan B.da Costa, Mozart N, Ramos, Rui Fausto. *Vibration spectra and structure of the cis and trans conformers of methyl nitrite: an ab initio MO study. Journal of Molecular Structure*,1996 . Vol.375. Pp.153-180.

[7]. Shaikhullina R M. , Khrapkovskii G M, Zverev E E, Mazilov E A *Quantum-chemical study of the molecular structure and vibration spectra of methyl nitrite*. Butlerov readings, 2010 . Vol.19 . №3.Pp.41-51 .

[8]. Bock Sh W, Krasnoshchiokov S V, Khristenko L V, Panchenko YN, *Pentin Y.A. Ab initio analysis of structure and vibration spectrum of methyl nitrate.* Chem.Phys. 106. 1985 . Pp.69-73.
[9]. Korolevich MV, Sivchyk V V, Zhbankov R G, Lastochkina V A *Theoretical and experimental study of frequencies and absolute intensities of the bands in the IR spectrum of methyl nitrate.* 1986 Journal of Applied SpectroscopyT.XLV. №6.Pp.945-951.
[10]. Muhamadeyeva R M (Shaikhullina), Zhbankov R G, Sopin V F, *Marchenko G.N. Low-frequency IR spectroscopy and the structure of cellulose and cellulose nitrates.* 1990 Minsk, Preprint, № 599.

New approach to investigating plasma processes occurring in friction units of machines and mechanisms

S Sharifullin and N Rogozhkin
Kazan (Volga region) Federal University, Kazan, Russia
E-mail: Saidchist@mail.ru

Abstract. Under the low temperature plasma usually we realize discharge plasma with a temperature of up to 50 000 °C. It turns out there is a discharges, which at first glance do not manifest themselves immediately, but play a crucial role in the creation of competitive vehicles, machinery and equipment with increased service life.

1. Introduction

Current national technical policy aimed at creating a competitive machinery and equipment with increased service life. However, no matter what technical solutions are not applied in engineering development, finite resource of the machine linked to the degree of surface wear-STAY its friction units. Depreciation is the result of friction. The process of friction can not be eliminated, ie. A. It is associated with the movement of the body and of course the conversion of mechanical energy into other forms of it. But to a certain extent reduce the effect of wearing out, you can. Or, without changing the friction process, you can improve the wear resistance of the material of the friction surface. Currently, there are so called resource conservation technologies (RES-technology), allowing not only extend the life of the machine, but also to ensure its without the wear operation.

The essence RES technology is that when administered to an established oil or grease composition of special machines called RVS additive on the surfaces of the friction units under the influence of pressure and temperature conditions are created for the return of wear on the worn seat surface with partial or complete restoration of the geometric dimensions of parts. Instead, wear the reverse process. Wherein at the friction surfaces are formed of layers with a high wear resistance and low friction coefficient [1].

Despite the fact that the first results of RES technologies have been obtained in the 80-ies of the last century to the present day it has not received widespread introduction into production. There are several reasons. One of them is the lack of a clear understanding of the mechanism of formation on the surface of the friction unit coating with high wear resistance and low friction. In recent decades, there have been some studies that reveal the essence of the processes in the friction of machinery in the presence of additives in them PBC. It was found that in these processes play a major role in the emerging contact patches mating surfaces friction unit electrical discharges. The question arises. From these ranks, and what role they are given in RES-technologies?

Sparking on impact and friction have long known. Only a few delved into its nature. In such technique called friction sparks sparks. They are the result of the transition of the mechanical energy into heat. When the relative sliding of two bodies due to the deformation of the material involved in the formation and destruction of frictional bonds in a thin surface layer formed by heat. Increased temperature can thus cause local softening and melting the material. Heat spreads from contact spots deep both contacting bodies, and heat flows are allocated depending on the thermophysical properties of the contacting bodies, the size of the heat sink and the conditions [2]. At steady-state thermal regime mechanical energy converted in micro surface layer into heat is transferred into the interior of the mass of the rubbing bodies by conduction from complete. Part of it is the result of structural transformations in the surface layers, thermionic and thermionic emission, thermal diffusion, and so on. D. Dissipated directly into the environment [2]. When a rough surface slides on the other, the touch occurs only at certain points which are quite a short time in contact. During the existence of a single spot can be 10-7 - 10-8 s [3]. Single spot on a short-term single temperature flash. Flashpoint can reach considerable size, which affects the conditions of interaction of the rubbing surfaces. According to the beliefs of the German physical chemist P.A. Thieβen, blow each other grains during grinding leads to the concentration of energy in the microscopic surface area (1967). As a result, at very short time intervals, about a thin layer of melt, and even the substance in Detects high-cal state, similar to plasma or magma. This "MAGMA-plasma" vposledstii Wii-called "Triboplazmoy" (Fig. 1 and 2).

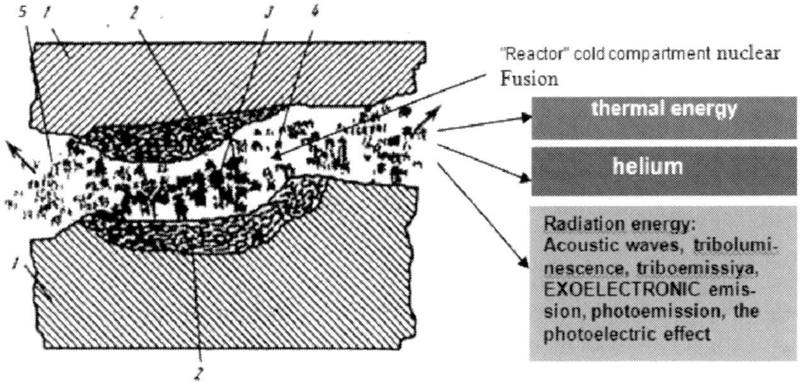

Fig. 1. Model of "Magma-plasma"

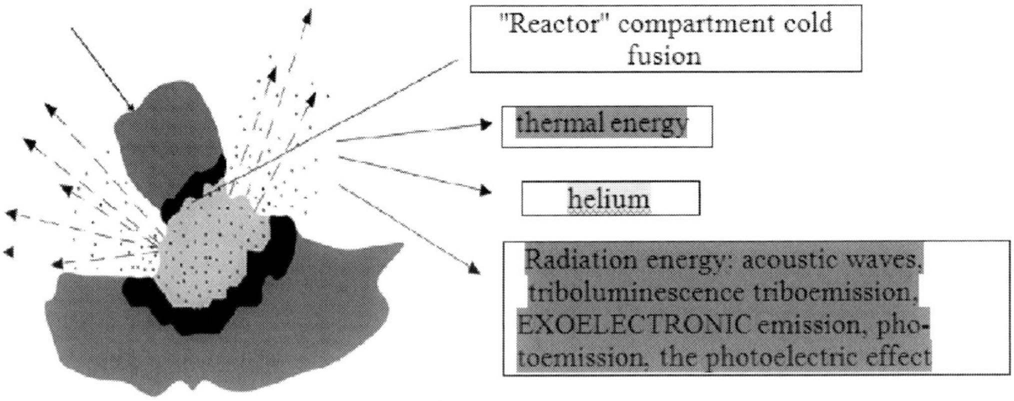

1 - the original structure; 2- molten structure; 3 - plasma; 4 - thermionic emission electrons 5 - atoms, photons, phonons, ions, excited molecules, the fast electrons.

Fig. 2. Model of "Magma-plasma" in the zone of impact of irregularities in surfaces. The impact direction shown by arrow

In spring 2008, Emeritus Professor Yoshiki Arata (Yoshiaki Arata) from the University of Osaka, and his Chinese counterpart Zhang Yuechan (Yue-Chang Zhang) from Shanghai University demonstrated public beaut-st experiment. In this demonstration experiment was recorded not provided the known laws of physics, the energy release and the formation of gen-

Leah [4]. In numerous articles devoted to the description of cold fusion (fusion at room temperature), focusing in particular on the fact that in experiments with deuterated substances heat is recorded neutron emission, tritium and helium.

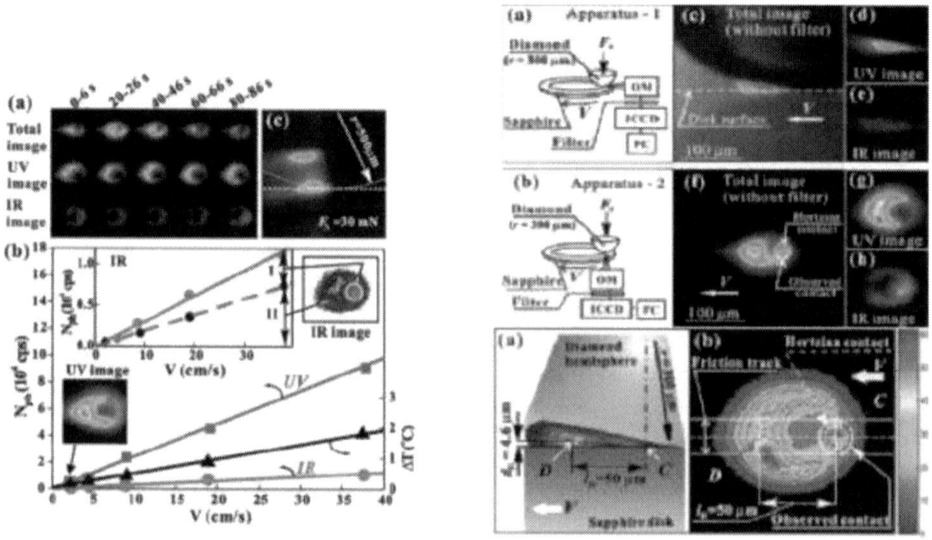

Fig. 3. Photos triboplazmy generated sliding contact between the diamond-dimensional sphere and a sapphire disk at atmospheric pressure air

According to the model, "Triboplazmy " released by mechanical vozdeyst-Wii on matter energy can significantly exceed the heat of fusion and vsledst-Vie low thermal conductivity of solids not only leads to the melting of the substance, but also to its sublimation, and then rise to a state in wherein it is in the form of ions and electrons (plasma state). For such applications, the laws of classical thermodynamics are not met; a thin surface layer of material is transformed, resulting in the collision zone, the irregularities are not formed plasma (triboplazma); process is accompanied by the emission of electron.

Keiji Nakayama in 2002 experimentally confirmed the presence triboplazmy around the sliding contact of the diamond sapphire sphere by rotating the disc (Fig. 3). Intense infrared radiation shows that the friction surface is heated. On the other hand intense ultraviolet radiation of the gas discharge observed outside the sliding contact between the localization of which has per-zorom sphere and disc ranging from 1.3 μm and 7.8 μm.

The observed plasma emits triboplazmu in sliding contact friction caused by heat or high energy formed in the deformed layer for sliding contact. Besides the emission intensity around how-zyaschego contact increases with increasing resistivity materials. This go-way indicates that that triboplazma electrification generated by friction.

The basis of the approach to the description of the structural features and behavior of matter tribomating in a state can be considered triboplazmy ideas about abnormally enhances the process of diffusion mass transfer in friction associated with changes covering the structure of the surface layerev tribocoupling [6]. At the same time observed as transformation has been accompanied by defects in the structure, and the reverse. On the phenomenon of surface flaw detection impose special imprint thermal processes, in particular, the diffusion coefficients, coefficient D0 corresponding triboplazme can be calculated using the equation:

$$D_0 = \frac{R_D^{\;2}}{h}(Q^* - W_\partial);\ [D_0] = \left[\frac{m^2}{c}\right]$$

$R_д$ – Debye radius, m;
h – Planck constant, J·S;
Q * = kT + Q – quantity characterizing the kinetic energy of thermal motion of the particles triboplazmy, J;
k – Boltzmann constant, J / K;
T – Temperature triboplazmy, K;
Q – heat released by friction, J;
W_∂ – activation energy for diffusion processes, J.

From this relation it follows that thermal activation surface greatly increases the activation energy of diffusion processes and leads to ano-mizing the diffusion coefficient. In further calculations we can obtain in the expression that reflects the proportionality of the diffusion coefficient of the square root of temperature field of frictional contact. From this it can be assumed that triboplazma their properties close to gases.

2. Conclusions

1. An electric discharge that occurs in friction machines and equipment, refers to the section of low-temperature plasma.
2. Triboplazma can be described by equations, applied to the Low Temperature-temperature plasma.
3. Research triboplazmy is relevant not only in terms of on-uchnogo interest, but also in terms of practical significance.

Acknowledgements
This work was funded by the subsidy of the Russian Government to support the Program of Competitive Growth of Kazan Federal University among World's Leading Academic Centers.

References

[1] Anatoly Dounaev. Friction Surfaces Modification Using Tribo-Compounds /Anatoly Dounaev and Said Sharifullin // World Applied Sciences Journal. ISSN 1818-4952, 2014. - Vol. 31 (2). - P. 272 - 276.
[2] KRAGELSKY IV Friction and wear. Ed. 2nd Revised. and ext. /I.V. KRAGELSKY. - M .: Mechanical Engineering, 1968 - 480s.
[3] Garkunov D.N. Tribotechnics /D.N. Garkunov. - M .: Mechanical Engineering, 1985 - 424c.
[4] Arata Y., and Zhang Y.-C. Formation of condensed metallic deuterium lattice and nuclear fusion. Proceedings of the Japan Academy. Ser. B: Physical and Biological Sciences. 2002. - Vol. 78. - No. 3. - P. 57 - 62).
[5] Tribology of Diamond-Like Carbon Films. Fundamentals and Applications. Christophe Donnet, Ali Erdemir Editors. (Christophe Donnet, University Institute of France and Uni-versity Jean Monnet, Laboratoire Hubert Curien. UMR 5513 18th avenue; Professeur Benoît Lauras 42000 Saint-Etienne, France. Donnet@univ-st-etienne.fr.
[6] Dolgopolov K.N. Improving the performance of household machines through intensification processes lubrication tribocoupling: Author. Dis. on soisk. Uch. steppes. cand. tehn. Sciences. / Dolgopolov Kirill Nikolaevich. SRSUES. - Mines, 2009. - 28 p.

Connection between machines durability during operation and the complex of physic-mechanical properties, formed during the production of critical parts

VG Shibakov, D L Pankratov, R V Shibakov
Kazan Federal University,
423800, 68/19 Mira avenue, Naberezhnye Chelny, Russia

E-mail: chelny@kpfu.ru

Abstract. The paper presents the effect of the processing technological structure on the hardening of metals and alloys by choosing the rational parameters of thermomechanical forming components at various stages of processing. The dependencies that can predict the destruction of the product during operation, depending on the structure oi the resulting metal fabrication are shown.

1. Introduction

Among mechanical engineering resources consumed metals and alloys play a special role by virtue of the inherent complex of unique physic-chemical and technological properties that constitute a constructive foundation of modern machines and are the subject of processing at the main stages of their production. Metals constitute 96% of a total aggregate of structural materials used (including plastics, composites, etc.).

Improving the metal quality leads to economic effects in different sectors, where the metal is transformed into structures and in their operation, due to the increased performance and technical features of a machine made of it.

2. Basic part

By virtue of the relationship and interdependence of different stages of the life cycle (see Fig. 1) , increased efficiency - reducing costs while operating the machine, due to the improvement of its technical parameters (power, durability, reliability , etc.) often results in costs increase in production, as the improvement of these parameters is often associated with the introduction of additional strengthening heat treatment operations or application of high expensive construction materials.

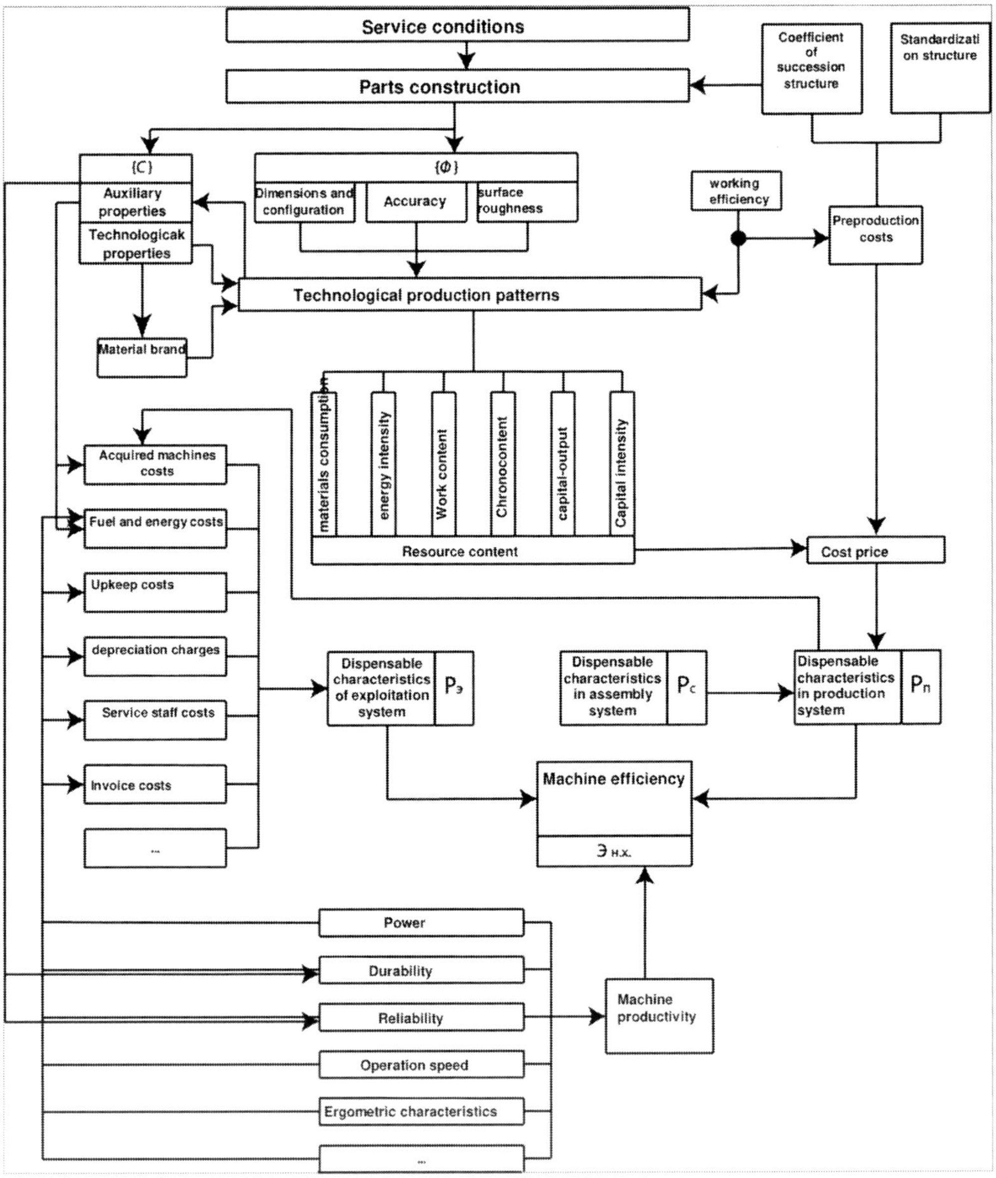

Fig. 1 Connection (relationship) of structural materials service and technological properties with parameters of durability and efficiency of transport vehicles

Nevertheless, the increase in costs in the production system, aimed at improving the service properties of structural materials , and consequently, an increase in durability , even leading to an increase in prices , it is advisable , since gains from increased durability far exceeds the reduction in economic effect due to increased cost of an automobile . Thus, an increase in durability relative to the average value of 3 times, followed by twice the automobile costs rise also 3 times increases the production economic effect [1].

Moreover the use of all technological possibilities for metals and alloys hardening by means of the rational choice of parts forming thermomechanical parameters at different stages of processing operation allows improving durability without introducing additional heat treatment that further enhances the economic efficiency associated with machines operation and connected with service properties of constructional materials used.
The connection of structural materials service and technological properties with the durability and efficiency of transport vehicles is displayed in Figure 1.

Durability of machine parts during operation depends on the processes occurring in the metal in operation and leading ultimately to the sudden or gradual failures. The greatest danger is sudden failures of parts, which can lead to a crash and serious damage to the whole machine. The destruction of deformation (ductile fracture) , brittle and fatigue damage leads to the sudden failure. Each type of failure other than structural and operational factors, depends on the parameters of the structure and mechanical properties of the metal parts.

One of the reasons causing the destruction of parts during operation, is the aging of the metal [2; 3; 4, 5, 6].

In the first stages of aging (the appearance of submicroscopic inhomogeneities in the distribution of solute atoms in a supersaturated solid solution, the formation of dispersed particles and the establishment of a coherent communication grids) occurs strengthening of the alloy. The final aging step leads to coagulation of precipitated dispersed particles, the loss of new phase gratings and solid solution coherence, the solid solution depletion during the extraction process and results in the alloy destruction. The alloy strength changing at this step is described by the equation [4]:

$$\ln\sigma = \ln\sigma k - k''te^{-Q/R},$$

where σk-the strength of alloy after hardening treatment and artificial aging (initial strength); σ - the current value of strength , k ''- integral factor comprising constants k and k ', depending on the nature of the material and component concentration ; t - aging time ; Q - activation energy , R - energy of the reaction . On the assumption of the equation the strength decreases with time exponentially with the external environment characteristics constancy. Until a certain moment the σ part strength will be sufficient to withstand the attached load L. It will be a period of no-failure operation. However, due to the aging of the material , after some period t, the strength is reduced to the level that does not conform to the load L, leading to the destruction (failure) . :

Knowing the initial strength and the load the material durability, and therefore part durability, taking into account aging equation can be calculated from the following relationship:

$$t = 1/\,k''(\ln\sigma - \ln L)e^{-Q/R}$$

Product life improvement due to the material can be achieved by its initial strength increase and the softening aging rate reduction, as well as removing in it the macrodefects of critical dimension. The latter can cause brittle fracture already at relatively low operating stresses. The adverse tensile technological residual stresses as a rule have a significant impact on the brittle fracture occurrence. The condition of brittle fracture, expressed in the form of Hall-Petch equation has the following view:

$$\sigma_b = k_B d^{-1/2}$$

Another form of the condition is the equation of the Griffith-Orowan-Irwin:

$$\sigma_b = \sqrt{2\gamma_p d},$$

where E - Young's modulus; - fracture surfaces effective energy; $k_B = \sqrt{2E\gamma_p}$; d - the size of the structural component (grain).

The part ability to carry high load items at the lowest possible weight and deformations most fully characterized by generalized index σ02E/j2, where j - density of the material [1].

Durability of parts under cyclic loading is characterized by the coefficient [8]:

$$K_{ув} = \sigma_{-1}/\sigma_{экв}$$

where σ_{-1} - endurance limit at a symmetric cycle ; $\sigma_{экв}$ - equivalent stress , depending on the type of load and stress state circuit parts during operation. By-turn σ_{-1} depends on σ_{02} and σ_B.$\sigma_{-1} = (0{,}2\div0{,}3)\sigma_B(1 + \sigma_{02}/\sigma_B)$, but σ_B is related to the d -metalla already given in the Hall-Petch relationship . Based on these relations it follows that for each type of fracture , causing failure, you can define the structure and related mechanical properties of the indicating parts durability during operation.

Consider the technological aspects of the formation of machine parts service properties. According to the previously given relations, parts durability stipulated by fracture failure depends primarily on the strength properties of the metal.

Strength is defined as the ability of a metal to resist irreversible (plastic) deformation when subjected to a load . Modern ideas of plastic deformation associated with its dislocations mobility. The resistance to deformation improvement can be achieved by means of the metal fine structure certain formation, retarding the motion of dislocations. Basic mechanisms of dislocations retardation are as follows:

- The formation of atoms aggregates (segregation) of alloying elements (openings) around dislocations in solid solutions ;
- Increase in the dislocation density, leading to atoms interaction consolidation of near the moving dislocations. In this case, the stress field of atom interaction forces interferes in the movement of some others;
- The formation of barriers to moving dislocations in the form of interface in crystals or particles of the second hardening phase of, ie creation of objects inside the alloy with different crystallography of dislocation slip .

The technological parts processing route has a decisive influence on the formation of optimal structural states of the metal and the corresponding strength parameters. Schematic diagram of the formation of service properties in the system of the metal into the part transformation is shown in Fig. 2, where

Su - basic properties of the material ; Sc - service properties of parts; a, d- processing routes with indirect influence of plastic forming technology (PFT) to Sc; c , b - processing routes with direct influence of PFT on Sc ; d, e - processing routes without influence TPF on Sc; N - normalization ; TI – thermo-improvement ; RA – recrystallization annealing ; DA - diffused annealing ; SpA - spheroidizing annealing ; SA - softening annealing ; PSA - pearlite structure annealing ; LGA - large grains annealing ; QT - quenching + tempering ; ChHT - chemical heat treatment ; FO - physical treatment (laser, plasmic , etc.); ThCT - thermo cyclic treatment ; CF - cold forging ; LTMTF and HTMTF - stamping in low-and high-temperature thermomechanical treatment mode; HSS- high speed stamping; SES – stamping in superelastic conditions ; HF - hot forging ; IThF - isothermal forging . [7].

Change metal properties in the process of transforming it into a piece should be analyzed by taking into account the technological heredity. Metal structure and properties depend not only on the finishing, but the features of all the preceding processes starting with the step of melting the metal on the operating step 1 up to acquisition workpieces on steps 2, 3, 4, 5.

Carriers of hereditary information are chemical composition, material structure and connected with it mechanical and other properties. To achieve high performance of metal components properties it is necessary to take into consideration or better – to manage technological inheritance. Those properties, which positively affecting the quality of parts, must be maintained and developed, as for negative characteristics, it is desirable to eliminate them at the initial stages of treatment.

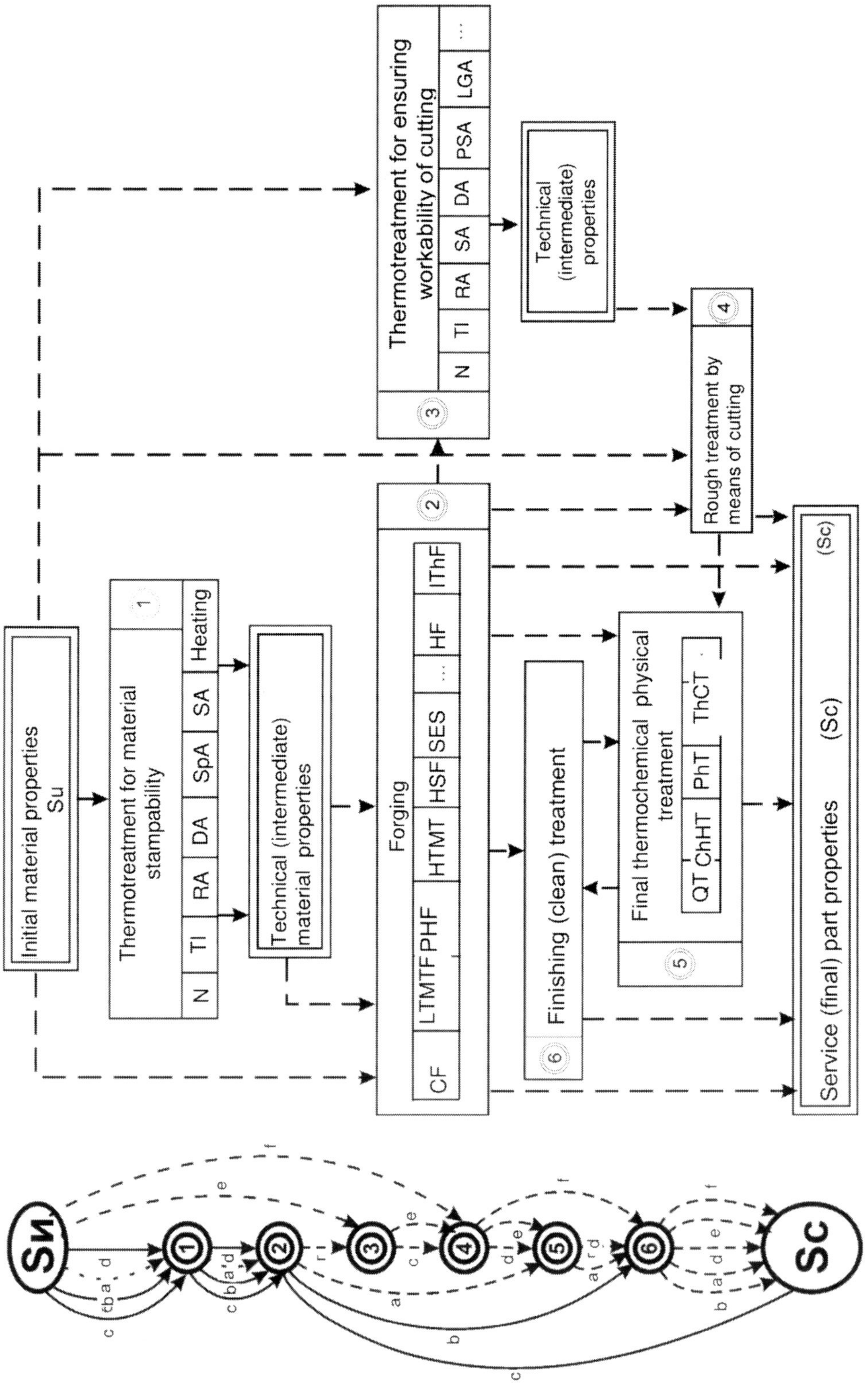
Initial material properties Su
Thermotreatment for material stampability
N
TI
RA
DA
SpA
SA
Heating
Technical (intermediate) material properties
Forging
CF
LTMTF
PHF
HTMT
HSF
SES
...
HF
IThF
Thermotreatment for ensuring workability of cutting
N
TI
RA
SA
DA
PSA
LGA
...
Technical (intermediate) properties
Rough treatment by means of cutting
Finishing (clean) treatment
Final thermochemical physical treatment
QT
ChHT
PhT
ThCT
Service (final) part properties (Sc)
(Sc)
Sи
Sc

Fig. 2. The scheme of part material technological and service properties forming in processing

Influence of certain technological treatment route on the metal properties change should be assessed in terms of inheritance coefficients [9]:

$$K_c = \frac{c_{i+1}}{c_i},$$

Where C_i, C_{i+1} - respectively, the values of some properties before and after exposure to the metal. Inheritance coefficients characterize changes in the properties at this stage of processing their values relative to the preceding stage. The value of K_c = I indicates that the *i-e* property is inherited without change, when Kc>I – property value increases, and when Kc<I decreases.

Table 1

Comparison of the mechanical properties of the parts produced by different technological routes (material steel 40X)

Type of aggregative technological [4]	Type of semi finished work piece for producing a detail	Mechanical properties of parts					
		σ^{-1}, MPascal	σ_{02}, MPascal	σ_B, MPascal	KCV, kilojoule/$м^2$	δ, %	Ψ, %
BI	Molding in sand-clayey mould	186 / I	290 / I	62 / I	56 / I	II / I	20 / I
BII	Profiled iron (normalization)	30 / 1,	320 / 1,1	69 / 1,1	75 / 1,35	20 / 1,8	
BIII	Forging from a cast form (normalization)	260 / 1,4	320 / 1,1	685 / 1,1	75,5 / 1,37	17,6 / 1,6	36 / 1,8
BIV	Forging from profiled iron (normalization)	379 / 2,03	455 / 1,17	866 / 1,13	83 / 1,5	2 / [illegible]	54 / 2,7

Ci - the numerator - the average absolute value of the metal parts properties;
K - the denominator - the coefficient of inheritance with respect to the properties of the metal part Ci obtained by crystallization of the melt in the sand-clayey mould.

$$K = \frac{C_2}{C_1}$$

Parts manufacturing process route significantly affects the mechanical properties of the metal. Table 1 presents a comparison of mechanical properties of metal parts made according to different processing routes. Data are presented for widely used structural steel 40X and averaged with the anisotropy of properties.

A higher level of parts mechanical properties produced by TCB BII , III, IV with using plastic deformation depends on special influence of the latter on the specific micro-and macrostructure of cast metal. Properly set plastic deformation modes promote to eliminate porosity of cast structure; to weld

IOP Conf. Series: Materials Science and Engineering **69** (2014) 012040 doi:10.1088/1757-899X/69/1/012040

macrodefects like cracks , crushing and rational orientation of nonmetallic inclusions ; redistribution of liquidation zones, forming a fibrous texture and deformation structure, grinding of sub- and macrograin , carbide constituent breakage and carbide inhomogeneity decrease and the like . [10]

Higher values of the metal strengthening characteristics can be achieved by deforming the metal while producing parts in the modes of low (LTMTF) and high (HTMTF) - temperatures of thermomechanical treatment. Data on the effect of various kinds of thermal and thermomechanical treatment on the strength and plastic properties steel 30KhGSA are shown in Table 2 .
Modes 3 and 4 (Table 2) by its temperature and speed settings ensure the passage of hot deformation in a deformed austenite dynamic polygonization process. At the optimal crank-type presses of deformation speed mode is not provided, so the strengthning properties of the metal forgings produced with this type of equipment in HTMTF mode are lower (Mode 8, Table 2) than those which can potentially be achieved.

Metal hardness and strengthening characteristics increase after HTMTF complicates subsequent machining process of forgings. Therefore, applying HTMTF at die forging is advisable if, as a result of stamping is possible to obtain the finished part or final shaping can be accomplished in the processes (grinding, electrochemical machining, etc.), which are not associated with non-blade processing.

Table 2

Comparison of metal parts mechanical properties produced by plastic deformation at various forming and heat treatment modes (30 HGSA steel material)

№ п/п	Forming and heat treatment modes	Mechanical metal properties					
		σ^{-1}, MPascal	$\sigma_{02,}$ MPascal	$\sigma_{B.}$ MPascal	KCV, Kjoule/м^2	σ, %	Ψ, %
1	Cold deformation (relational deformation 60%)	560 / 1,7	1100 / 1,8	1200 / 1,54	75 / 0,95	3 / 0,48	18 / 1,1
2	Cold deformation + tempering 300C° + high speed annealing + tempering 300 C° (preliminary thermal treatment PTMT)	1200 / 3,6	1640 / 2,6	1860 / 2,42	200 / 3,57	9 / 1,31	30 / 1,9
3	Half-hot deformation at low temperature thermomechanical treatment mode (LTMTF)	1890 / 5,6	2200 / 3,5	2700 / 3,49	400 / 5,14	7 / 1,02	25 / 1,6
4	Hot deformation at high-temperature thermomechanical treatment mode (HTMTF)	1760 / 5,32	2000 / 3,2	2400 / 3,12	500 / 6,4	10 / 1,46	35 / 2,2
5	Hot deformation + hardening with forming temperature + tempering 300C° (TMTF)	1490 / 4,2	1680 / 2,7	1910 / 2,43	250 / 3,2	10 / 1,46	30 / 1,9
6	Hot deformation + thermo-improving	930 / 2,8	1410 / 2,3	1690 / 2,2	185 / 2,38	10 / 1,46	25 / 1,6
7	Hot deformation + normalization	450 / 1,34	650 / 1,1	850 / 1,1	116 / 1,5	13 / 1,9	30 / 2,4
8	Hot deformation + - thermo cyclic treatment 14-15 cycles (ThCT)	409 / 1,2	550 / 0,92	815 / 1,05	314 / 4,05	21 / 3,0	62 / 4

IOP Conf. Series: Materials Science and Engineering **69** (2014) 012040 doi:10.1088/1757-899X/69/1/012040

Mechanical properties are averaged taking into account the anisotropy: in the numerator - the absolute values of the properties, the denominator - inheritance coefficient K after thermomechanical treatment.

Thus, the choice of a rational thermomechanical plastic deformation mode can purposefully change the properties of components, improving their service characteristics, particularly durability.

3. References

[1. Orlov PI *Design Principles*.1977 M. : Engineering Vol. 3, - 663 p.
[2] . Dmitriev A., Pankratov D L, Shibakov VG *Restoring parts by plastic deformation (theory, technology, equipment)* 2012 Moscow : Academia, 387p.
[3]. Bogdanoff J, Kozin F. *Probabilistic models of damage accumulation : Translated from English.* 1989 M.: Mir, 344 p.
[4] . Melamedov I M *Physical basis of reliability*. 1970 L. : Energy , 152 p.
[5]. Galyautdinov R T and Kashapov N F 2001 Iz. VUZ. *Aviatsionnaya Tekhnika* **2** 80.
[6]. Abdullin I Sh, Kashapov N F and Kudinov V V 2000 *Physics and Chemestry* **3** 35-38.
[7]. Orlov P I *Design Principles* 1977 M. : Engineering , Vol. 3, - 663 p.
[8]. Gurevich S E *Metal strength at cyclic loads* 1967 Moscow: Nauka , 360 p.
[9]. Rauzin Y R, Shur E A *The structural strength of the steel.* 1975 M . : Mechanical Engineering 56
[10]. Tyurin V A *Theory and processes of ingots forging on processes*. 1979 M . : Mechanical , 240 p.

Innovative Mechanical Engineering Technologies, Equipment and Materials-2013 IOP Publishing
IOP Conf. Series: Materials Science and Engineering **69** (2014) 012041 doi:10.1088/1757-899X/69/1/012041

The determining of the coefficient of safety of bearing ability of anisotropic bars in the general case of their complex resistance

K E Sibgatullin, E S Sibgatullin
Kazan Federal University,
423800, 68/19 Mira avenue, Naberezhnye Chelny, Russia

E-mail: chelny@kpfu.ru

Abstract. The bars of any form made of a uniform anisotropic material are considered. Generally in the cross section of a bar all internal power factors (IPF) – three forces and three moments are other than zero. Values IPF are known from the solution of the corresponding task. The coefficient of a stock of bearing ability of a bar is defined by a way of comparison of known vector IPF $\overrightarrow{R^*}$ with the corresponding required vector of durability $\vec{R}$ in IPF space.

1. Introduction

In many science courses on resistance of materials [1-9], as a rule, special cases of complex resistance of timber are considered: oblique bending, eccentric compression (extension), bending with torsion. The influence of shear forces on the destruction of the bar under its complex resistance is often neglected, considering this influence as insignificant. As a rule, checking the strength of the bar is being conducted, comparing the greatest normal (or equivalent) voltage $\max|\sigma|$, operating in a dangerous point of timber, with dangerous proper voltage $|\sigma_u|$. In this article the technique of checking strength of anisotropic bars of arbitrary shape in the most general case of complex resistance is given, while remaining in the IPF space. The article is a continuation of work [10].

2. Basic part

Fig.1 shows the IPF of the effective cross-section bar: T_{11} – normal force; Q_{21}, Q_{31} – cross-axis forces; M_1 – torsion torque; M_2, M_3 – bending torques.

We believe that the vector of the IPF $\overrightarrow{R^*}\{T_{11}^*, Q_{21}^*, Q_{31}^*, M_1^*, M_2^*, M_3^*\}$ is known. The corresponding vector strength $\overrightarrow{R}\ \{T_{11}, Q_{21}, Q_{31}, M_1, M_2, M_3\}$ and a factor of the bearing capacity of the considered section $k = \left|\vec{R}\right| / \left|\overrightarrow{R^*}\right|$ must be defined (1).

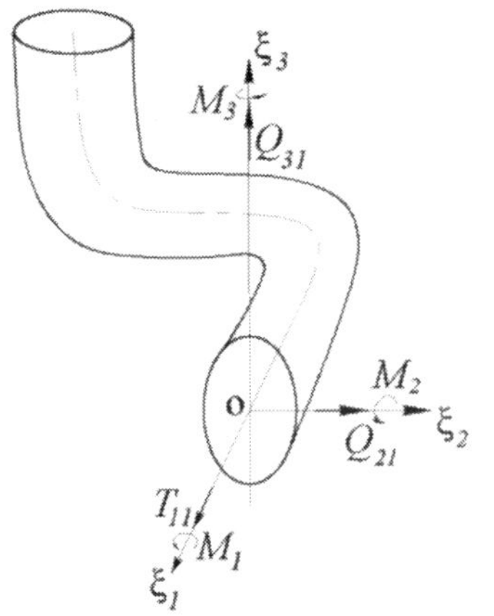

Fig. 1. IPF of the effective cross-section bar

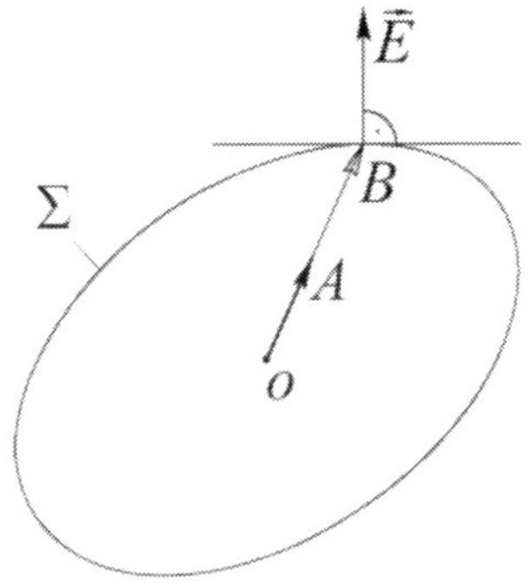

Fig.2. Limiting surface Σ in IPF space

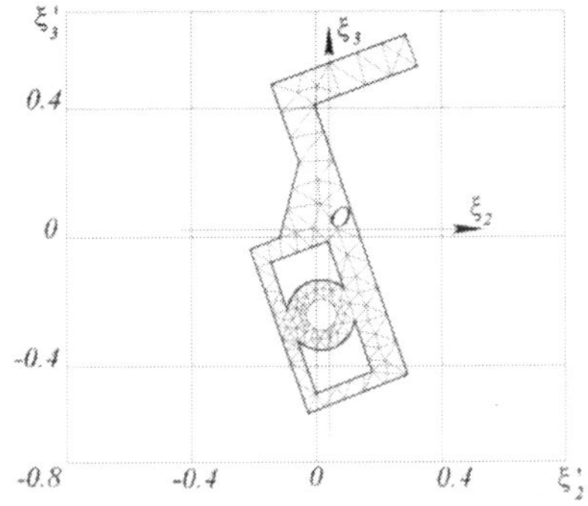

Fig.3. Cross-section of a bar

IOP Conf. Series: Materials Science and Engineering **69** (2014) 012041 doi:10.1088/1757-899X/69/1/012041

On Fig. 2 the limiting surface Σ in IPF space (surface strength) is schematically shown where $\overrightarrow{OA} = \overrightarrow{R^*}$ is a known vector of IPF; $\overrightarrow{OB} = \vec{R}$ – the sought vector strength corresponding to the vector $\overrightarrow{R^*}$. It is obviously, that vectors $\overrightarrow{R^*}$ and $\vec{R}$ should lie on the same straight and be equally directed; their origin should coincide with the beginning of the space coordinates IPF.

Let the surface of strength for anisotropic material in space voltages $\sigma_{ij}(i, j = \overline{1,3})$ be set by the equation

$$\Phi(\sigma_{ij}) \equiv A\sigma_{11}^2 + L\sigma_{12}^2 + K\sigma_{13}^2 + 2P\sigma_{11}\sigma_{12} + 2S\sigma_{12}\sigma_{13} + 2G\sigma_{13}\sigma_{11} +$$
$$+ 2D\sigma_{11} + 2Q\sigma_{12} + 2T\sigma_{13} - 1 = 0$$

(2)

In [10], based on the equations (2) and using the rigid-plastic model of deformable rigid body, the following parametric equations limit surface in the space IPF for anisotropic bars were received:

$$T_{11} = \frac{1}{2\Delta}[(\delta_{11}\dot{e}_{11} + 0{,}5\delta_{12}\dot{\gamma}_{21} + 0{,}5\delta_{13}\dot{\gamma}_{31})I_1 + (0{,}5\delta_{13}\dot{æ}_{11} - \delta_{11}\dot{æ}_{21})I_2 +$$
$$+ (\delta_{11}\dot{æ}_{31} - 0{,}5\delta_{12}\dot{æ}_{11})I_3] - \frac{\Delta_1}{\Delta}A_*; \quad \ldots;$$

(3)

$$M_3 = -\frac{1}{2\Delta}[(\delta_{11}\dot{e}_{11} + 0{,}5\delta_{12}\dot{\gamma}_{21} + 0{,}5\delta_{13}\dot{\gamma}_{31})I_2 + 0{,}5(I_4\delta_{13} - I_5\delta_{12})\dot{æ}_{11} -$$
$$- I_4\delta_{11}\dot{æ}_{21} + I_5\delta_{11}\dot{æ}_{31}] - \frac{\Delta_1}{\Delta}S_3$$

where $I_1 \div I_6$, S_1, S_2 are integrals on square A_* of cross-section of a bar; $\Delta, \Delta_i, \delta_{ik} = \delta_{ki}(i,k = \overline{1,3})$ – quantities dependent on the equation coefficient (2). Speeds $\dot{e}_{11},\ldots,\dot{æ}_{21}$ are in integrals $I_1 \div I_6$ [9].

The parameters in equations (3) are the relationships of the velocity vector components of generalized displacements $\vec{E}\{\dot{e}_{11}, \dot{\gamma}_{21}, \dot{\gamma}_{31}, \dot{æ}_{11}, \dot{æ}_{31}, \dot{æ}_{21}\}$. In general it is impossible to exclude these options from equations (3), using the exact mathematical methods. Thus, the practical application of the equations (3) is inseparably linked with the use of the computer for calculations. On the given vector of generalized velocities of displacement, using equations (3), it is relatively easy to find the corresponding vector strength (for example, vectors $\vec{E}$ and $\overrightarrow{OB}$).

To have an idea about the size and shape of the marginal surface (3), it is to be built on the points of its typical cross-section. These sections may be helpful in determining the safety factor for the bearing capacity of the bars in some cases, of their complex resistance. The range of practical applications of equations (3) would be significantly wider than if it is possible to use them in determining the stock carrying capacity of anisotropic beams in the most general case it difficult resistance. The below method is suggested for solving this problem.

Components of collinear vectors satisfy the following equations:

$$\frac{T_{11}}{T_{11}^*} = \frac{Q_{21}}{Q_{21}^*} = \frac{Q_{31}}{Q_{31}^*} = \frac{M_1}{M_1^*} = \frac{M_2}{M_2^*} = \frac{M_3}{M_3^*} = k \quad (4)$$

Using (4), the following system of equations can be got:

$$Q_{21}^{*}T_{11}-T_{11}^{*}Q_{21}=0;\ Q_{31}^{*}Q_{21}-Q_{21}^{*}Q_{31}=0;\ M_{1}^{*}Q_{31}-Q_{31}^{*}M_{1}=0;$$
$$M_{2}^{*}M_{1}-M_{1}^{*}M_{2}=0;\ M_{3}^{*}M_{2}-M_{2}^{*}M_{3}=0$$
(5)

Putting the results from (3) into (5), we get

$$Q_{21}^{*}\left\{\frac{1}{2\Delta}[(\delta_{11}\dot{e}_{11}+\ldots)I_1+\ldots-\frac{\Delta_1}{\Delta}A_*\right\}-T_{11}^{*}\left\{\frac{1}{2\Delta}[(\delta_{21}\dot{e}_{11}+\ldots)I_1+\ldots-\frac{\Delta_2}{\Delta}A_*\right\}=0;\ \ldots;$$
$$M_{3}^{*}\left\{\frac{1}{2\Delta}[(\delta_{11}\dot{e}_{11}+\ldots)I_3+\ldots-\frac{\Delta_1}{\Delta}S_2\right\}-M_{2}^{*}\left\{-\frac{1}{2\Delta}[(\delta_{11}\dot{e}_{11}+\ldots)I_2+\ldots-\frac{\Delta_1}{\Delta}S_3\right\}=0$$
(6)

Since in this case only the direction of the vector $\vec{E}$ is essential and its length may be arbitrary, as an additional to the system (6) we will adopt the following equation:

$$\dot{e}_{11}^2+\dot{\gamma}_{21}^2+\dot{\gamma}_{31}^2+\dot{æ}_{11}^2+\dot{æ}_{31}^2+\dot{æ}_{21}^2=1 \quad (7)$$

The algorithm for determining the factor of anisotropic bars is the following:

1°. To define a vector IFP $\overrightarrow{R^*}\{T_{11}^*,Q_{21}^*,Q_{31}^*,M_1^*,M_2^*,M_3^*\}$ in a dangerous section of the rod as a result of solution of the corresponding problem, for example, the methods of "Resistance of materials".

2°. Solving the set of equations (6), (7) relatively to find the vector associated with the vector $\vec{R}$ of the strength of the associated law [11] (Fig.2)

3°. Substituting found in paragraph 2 vector components in equation (3), to identify the components of the strength vector $\vec{R}\{T_{11},Q_{21},Q_{31},M_1,M_2,M_3\}$.

4°. To check compliance with the conditions (4).

5°. According to the formula (1) determine the factor k anisotropic bars by method of limiting conditions.

Some results of calculations. Take the bar cross section of which is shown on Fig. 3. Here ξ_2,ξ_3 are the main central axis of the square cross, ξ_2', ξ_3' - section-arbitrary axis, dimensions are given in meters. The strength properties of material of the bar is described by the equation

$$0.2\sigma_{11}^2+30\sigma_{12}^2+60\sigma_{13}^2-0.1\sigma_{11}\sigma_{12}-\sigma_{12}\sigma_{13}-0.2\sigma_{13}\sigma_{11}-$$
$$-0.6\sigma_{11}+14\sigma_{12}+20\sigma_{13}-1=0$$
(8)

Here the voltages are attributed to a characteristic value σ_0, having the dimension of stress. In [8] for the timber, using (8) and (3), cross section limit surfaces were built (3) by coordinate planes $T_{11}OM_1$, $T_{11}OM_2$, $T_{11}OM_3$, M_1OM_2, M_1OM_3, M_2OM_3, $Q_{21}OQ_{31}$.

The table below lists the results obtained for illustrated in Fig. 3 section using the above-proposed algorithm. Here $T_{11}^*,Q_{21}^*,Q_{31}^*,M_1^*,M_2^*,M_3^*$ are the known values of IFP.

$T_{11}, Q_{21}, Q_{31}, M_1, M_2, M_3$ - relevant components of the strength vector $\overrightarrow{R}$, defined by the decision of systems of the equations (6, 7).

Table

№	$\frac{T_{11}}{T_{11}^*}$	$\frac{Q_{21}}{Q_{21}^*}$	$\frac{Q_{31}}{Q_{31}^*}$	$\frac{M_1}{M_1^*}$	$\frac{M_2}{M_2^*}$	$\frac{M_3}{M_3^*}$	k
1	0.369 0.741	-0.0465 -0.0932	-0.0330 -0.0662	0.000538 0.00106	-0.308 -0.617	-0.00606 -0.0118	0.5
2	0.314 0.0699	-0.0466 -0.0103	-0.0331 -0.00737	-0.000560 -0.000124	0.296 0.0658	0.0343 0.00763	4.5

Summary conclusion. The proposed methodology strength check of anisotropic bars on limit states has, in our opinion, the following advantages:

1. The need for separate consideration of private kinds of complex resistance of timber (oblique bending, bending with torsion and the like) becomes unnecessary.

2. Each of the components of the vector $\overrightarrow{R^*}$ of IFP has equal rights to participate in the test of strength.

3. only the most general requirements accepted in the theory of rods are imposed on the form and the dimensions of timber.

4. The ratio $\left|\overrightarrow{R}\right| / \left|\overrightarrow{R^*}\right|$ is closer to reality assesses the robustness of timber than the ratio $|\sigma_u| / \max|\sigma|$ (this remark is especially true if the material of the bar has plastic properties). As noted in [12], "... calculation of limit loads gives significantly more correct understanding of the structural strength, than the calculation of the maximum stress...". The proposed methodology strength check the bars expands the sphere of practical application of equations (3); it can be assumed that it will find application in practice of calculation and design of various products.

3. References

[1] Ilyushin A A, Lenskiy V S 1959 *Resistance of materials*.M.: Fizmatgiz 372 p
[2] Rabotnov Yu N 1962 *Resistance of materials*.M.: Fizmatgiz. 456 s.
[3] Darkov A V, 1975 Shpiro G S *Resistance of materials*.M.: Vyissh. shk., 654 s.
[4] Smirnov A F, Aleksandrov A V, Monahov N I i dr. 1975 *Resistance of materials*.M.: Vyissh. shk 480 s
[5] Timoshenko S P, Gere D 1976 *Mechanics of materials*. M.: Mir 671 s
[6] Teregulov I G 1984 *Resistance of materials and the basic theory of elasticity and plasticity*.M.: Vyissh. shk 472s
[7] Pisarenko G S, Agarev V A, Kvitka A L i dr. 1986 *Resistance of materials* Kiev: Vischa shk 775s
[8] Abdullin I Sh., Kashapov N F and Kudinov V V 2000 *Physics and Chemistry of Materials Processing* **3** 35-38
[9] Abdullin I Sh, Kashapov N F and Kudinov V V 2000 *Physics and Chemistry of Materials Processing* **2** 40-44

[10] Sibgatullin K E, Sibgatullin E S 2010 *Evaluation of the strength of anisotropic beams with arbitrary cross-section in the general case of a complex impedanc*//Izv. RAN. MTT.. #1. S 84-92
[11] Kachanov L M 1969 *Fundamentals of the theory of plasticity* M.: Nauka 420 p
[12] Kachanov L M 1974 *Fundamentals of fracture mechanics* M.: Nauka 312 p

Innovative Mechanical Engineering Technologies, Equipment and Materials-2013 IOP Publishing
IOP Conf. Series: Materials Science and Engineering **69** (2014) 012042 doi:10.1088/1757-899X/69/1/012042

Modular representation of the product in the knowledge base in the technological process formation

L A Simonova, E I Egorova
Kazan Federal University,
423800, 68/19 Mira avenue, Naberezhnye Chelny, Russia

E-mail: chelny@kpfu.ru

Abstract. The article describes the stage of engineering preparation of production, where the knowledge base is offered in the form of a matrix, which will be an active information resource for a stage of technological pre-production.

1. Introduction
Modern technologies for polymeric products producing and processing include heat mass exchange Knowledge base provides a quantitative and qualitative analysis of the solution based on the internal algorithms and the aggregate of all the decisions taken once, forming the basis of precedents. In the database of precedents search, classification and analysis of the solutions storage is performed in a formalized manner [1-3,6,7].

2. Basic part
The system under consideration is used to support decision-making in a very complex area. Knowledge base containing a significant amount of decision rules is difficult for perception and use. It is therefore proposed to build the knowledge base in a modular fashion. Decomposition of knowledge bases is made in accordance with the hierarchical structure of the process.

Advantages of modular knowledge bases are the following:

- modules provide tools and techniques to maintain knowledge bases stored in separate files and use these files by multiple applications;
- modularization is taken into account by the mechanism of output in the process of work, which improves its efficiency;
- modules can be selectively activated or deactivated, allowing you to exclude from consideration the rules of inactive modules.

Applying the above said to the problem statement, we decompose, with the purpose of identifying the relationship of generic «*is a*».

Any technical engineering product is a system described by a set of elements, the attitude to which depends on what angle the product is being viewed. Once the designer has designed a product, it must be produced, and since then it should be viewed as an object of production. If the technical system is viewed as an object of production, it is a set of assembly units and parts, impersonal in terms of their functionality. Designer in the preparation of technical documentation for a product gives the information on the item as an object of exploitation. For a technologist it is important at the stage of getting the task to produce new products quickly and accurately, to determine the expected complexity

IOP Conf. Series: Materials Science and Engineering **69** (2014) 012042 doi:10.1088/1757-899X/69/1/012042

of manufacturing the product, the need for technological facilities and the possibility of producing it under the operating conditions. However, the information that is presented in the design documentation is not sufficient to solve this problem. At present, this information is, to a great extent, in an ***implicit form***, and for obtaining it a highly skilled technologist and much time are required, which is unacceptable from the viewpoint of the product lifecycle, especially at the stage of design and production planning [4].

The task would be greatly simplified if the information was contained in the form of active information resources, that is represented an explicit table, containing the information on all the compounds, sets of main and additional bases with levels of accuracy and roughness of their surfaces, levels of precision of the parts.

To build up the production process of a part, a technologist needs to know not so much the brand of the material, but as the material processing characteristics and the recommended methods of the material processing. In this regard, there is a need to develop a method of describing a product as an object of manufacture. If we consider the product as a set of its constituent parts surfaces, then to evaluate the production of an item without developing the process is possible by using the complexity of technological transitions in manufacturing ***surfaces of parts*** and ***the connection of the parts.***

In this regard, a product design should be presented as a set of connections and surfaces of parts. To do this, the product is considered as a structured set of connections and surfaces of parts. The surfaces of parts are such elementary geometric surfaces as plane, cylinder, cone, etc. However, a part performs its functions, using a combination of surfaces. Then it is necessary to find in the design of a product such an element, the production technology of which was an element of the process to be included into it without significant changes. These elements are a combination of parts' surfaces, due to which the part performs its functions called ***modules of surfaces***. Then, any part can be described by the modules of the surface [5].

Decomposition of classes and objects is brought to the level of elementary surfaces of parts, which are basic components of a complex system. Moreover, if we consider the levels of the hierarchy from top to bottom, the upper level is the class, such as a reducer. In relation to the upper level the following (sub) hierarchy level forms class objects, such as a framework, a shaft - gear, etc. The next lower level is an object for the higher level and the class level for a lower level, and so on up to the level of elementary surfaces.

When creating an intelligent system the fact, that there are much more objects in a complex system, than classes, is taken into account. Application of the hierarchical approach makes it unnecessary to repeat the same information for each example of an object, that is, with the introduction of the class structure it becomes possible to place in them the general properties of the examples. If this approach is used in the subject area, namely in engineering, all that engineers design, is designed to perform specific functions. The products of main and additional production, as well as related processes, have their functionality. Projects developed with the help of computer technology, are the objects that are seen by these resources. Then, during the production these projects are transformed into tangible objects that represent objects of production, which are perceived by human thinking.

Design and technological documentation consists of displaying the form of entities with certain content, which is the object [4]. All properties have some value that can be simple quantitative or qualitative (symbolic) characteristics, and may refer to another object. The sizes with deviations or maximum deviations; deviations of a form and a surface position; surface roughness values, etc. can be attributed to quantitative characteristics of products. Qualitative characteristics of products are the name and the brand of a material; fundamental size deviations, etc. Quantitative characteristics of technological operations include set-up time, the number of workers carrying out the operation, and qualitative characteristics – the code and the name of the operation, the code and the name of the equipment. To develop a product design or a process it is necessary to assign specific values to all the properties of the corresponding object and all the objects associated with the attitude "the whole – a part". In this way an example of an object is formed. A dynamic nature of the current properties of the object's values is associated with the frequent need to make changes. Consequently, the behavior of the object is shown in the structural - parametric synthesis of design and technology.

In the model of a module CAD intelligent system of building up process, classes and objects are related to the hierarchical structure. In the process of structural synthesis in the method of the involved object - class a decision about choosing one of the objects - subclasses is taken and values of their own properties are defined. In the result of the parametric synthesis the selected object - subclass is activated, and it inherits the properties. In this process, technical documents are received, which reflect the state and behavior of the designed objects.

Intelligent system of building processes is a complex system, the implementation of which comes in the form of object interaction. The conceptual interaction of the objects occurs as the links between them. The links can be structural and parametric. Structural links are defined by the hierarchy of objects given in the Meta system, and parametric ones can be vertical and horizontal. Moreover, the vertical links inherit the properties by hierarchy and the horizontal ones transfer property values between the objects, not subordinate to each other [6, 9, 10].

Thus, the output stage of the design pre-production will be a knowledge base in the form of a matrix, which will be an active information resource for technological production preparation stage.

3. References

[1] Simonova L A, Kostyuk I V 2009 *Formation of knowledge base system of tooling management Forging and stamping production* Metal Forming № 8 pp 22-28

[2] Simonova L A *Formation of knowledge database of alternative technological routes' parameters* (2004) Kuznechno-Shtampovochnoe Proizvodstvo (Obrabotka Metallov Davleniem) (10) pp 44-48

[3] Larisa A Simonova, Marat R Khisamutdinov 2013 *Intellectual Model Control Data of the Module Integration SAP-ERP and Teamcenter*-PLM. World Applied Sciences Journal 25 (8): pp 1258-1262

[4] Bazrov B M 2007 *Fundamentals of Manufacturing Engineering*: A Textbook for high schools. 2nd ed. Mashinostroyeniye 736 p il

[5] Norenkov I P, Kuzmik P K 2002 *Information support of high-tech products CALS-technologies* - M.: Publishing house MSTU. N E Bauman 320 p

[6] *Knowledge Base of Intelligent Systems* / T A Gavrilova, V F Horoshevsky - St. Peter 2000 384 p Il

[7] *Intelligent systems of design: tutorial* / G B Eugenev Moscow: Publishing House of the MSTU. N E Bauman 2009 334 [2] p Il

[8] Popov E V 1987 Expert *systems: the decision of unformalized problems in dialogue with the computer* Moscow: Nauka

[9] Simonova L A 2012 *Information Software of management of technology routes* (monograph) Germany: Publishing House Palmarium Academic Publishing 186 p

[10] Booch G 2000 *Object-Oriented Analysis and Design with the examples of applications in C + +:* Tr. from English. 2nd ed. M. Binom, St. Petersburg.: Nevsky Dialect

Cleaning of a surface of details in plasma of the volume category between solid and liquid electrodes

R I Valiev, B Yu Shakirov, Yu I Shakirov
Kazan Federal University,
423800, 68/19 Mira avenue, Naberezhnye Chelny, Russia

E-mail: chelny@kpfu.ru

Abstract. Plasma electrothermal installation with the liquid cathode is developed. Cleaning modes from a surface of products are investigated and fulfilled.

1. Introduction

One of the main technological operations at production and restoration of worn-out details in mechanical engineering is operation of cleaning of a surface. The operation of cleaning of the surface of semiconductors should be carried out in instrument making and electronics. Cleaning [1; 2, 3, 4] defines quality and reliability of operation of the unit, increases its operational properties, and for marketable appearance.

2. Experimental plasma electrothermal installation with a liquid electrode for cleaning from the surface of products

Experimental plasma electrothermal installation is developed for optimization of process of cleaning with a liquid electrode [5; 6; 7; 8] (fig. 1). The bathtub is made of copper plates and has a cooling shirt. It is filled with researches electrolytes of necessary concentration. The bathtub is connected to one of power supply plugs. The top solid electrode joins other plug of a source. This electrode is established on the coordinate device, allowing to regulate interelectrode distances and on X, Y coordinates. As electrolyte were used NaCl, $CuSO_4$, NH_4NO_3.

Experimental installation is intended for research of an electric discharge in the range of the parameters Up =0,3 – 3000V, currents of I=0,01÷200 A, $j_э$ =0,1 - 25 A/cm^2, interelectrode distances l = 0,1 – 100 mm. The cooling system of installation serves for electrolyte cooling. It comes from the high-level water supply system, and supply of electrolyte is carried out from a special tank.

Current passing through an electrolytic cell, in which the surface of anode is much less than at the cathode. At a small tension current passing in volume of solution is described by the Ohm law, and processes on electrodes is described by the Faraday's laws. The increase in tension leads to an electrolyte warming up mainly in an attached to anode zone as in it almost all resistance of an electrolytic cell is concentrated.

Transitions from one mode of passing of current to another happen through achievement of critical values of tension. For formation of a continuous steam-gas cover (fig. 2), at least unstable, it is necessary to provide allocation in an attached to anode zone of energy, sufficient for electrolyte boiling up in some volume.

For formation of a steam-gas cover it is necessary to heat the attached to anode layer of solution to a boiling point [9].

IOP Conf. Series: Materials Science and Engineering **69** (2014) 012043 doi:10.1088/1757-899X/69/1/012043

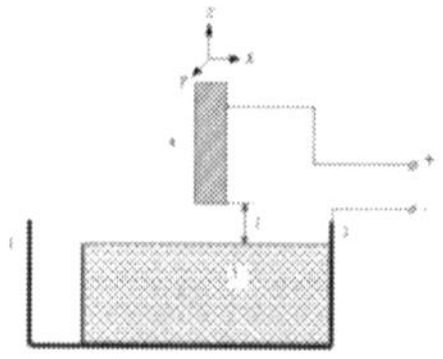

Fig. 1. The scheme of an electrolyti
the flowing electrolytic cathode
1 – electrolytic bathtub;
2 – current lead (metal bathtub);
3 – electrolyte;
4 – metal anode

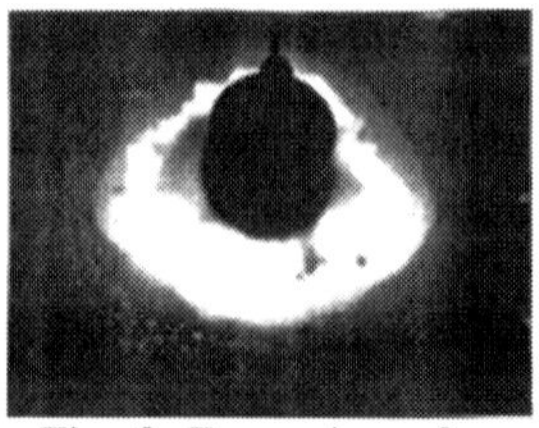

Fig. 2. Formation of a steam-gas cover the electric discharge, burning between electrolyte and a porous layer under atmospheric pressure. Depth of immersion of a detail in electrolyte m h=0,001.

Voltamper characteristics of a cell are written down by means of the two-coordinate recorder PDP-4-002 when giving tension increasing on the linear law with a constant speed in the range from 1 to 30 V /s. Voltamper characteristics are presented in fig. 3.

The received dependences were, as a rule, linear. In solutions of high concentration resistance of a cell is almost constantly in all range of tension (curves 1 and 2 in fig. 3).

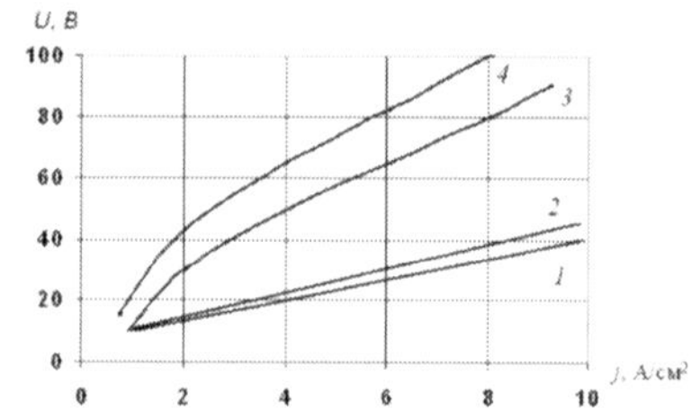

Fig. 3. Voltamper characteristics of a cylindrical cell. Concentration of NH_4NO_3 (%): 1, 2 - 25; 3, 4 - 5. Tension growth rate (V / s): 1 - 1,3; 2 - 10,9; 3 - 4,0; 4- 25,0

Main contents and result of work

The received voltamper characteristics and calculations allow to develop plasma electrothermal installation for cleaning from the surface of products.

We offer process of cleaning of the surface of semiconductors, non-ferrous metals and steels by means of the plasma electrothermal installation (fig. 4), burning between the liquid cathode – 20 ÷ 25% - water solution NaCl and the solid anode shipped in it - a cleared product, methods of removal of agnails and fair processing.

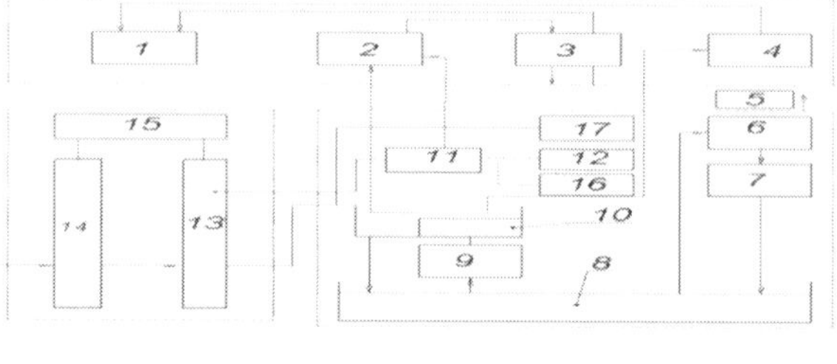

Fig. 4. The block diagram of plasma electrothermal installation with a liquid electrode

1 - voltage stabilizer; 2 - interelectrode gap monitoring system; 3 - monitoring system of parameters of electrolyte; 4 - monitoring system of electric parameters of the category; 5 - exhaust ventilation; 6 - pump of pumping of electrolyte; 7 - the filter for purification of electrolyte; 8 - capacity with electrolyte; 9 - the pump of supply of electrolyte in an electrolytic bathtub; 10 - electrolytic bathtub; 11 - processed detail; 12 - the coordinating device; 13 - rectifier block; 14 transformer; 15 - tension regulator; 16 - detail fastening; 17 – counter of amount of electricity.

In this case at U> U_k where U_k - cathodic power failure, burns the high-voltage category in formed on the electrolyte surface to the steam-gas environment. This category consists of a set of microcategories in the colloidal environment with porous capillary structure. Our supervision showed that these microcategories clear the surface of semiconductors for 20-30 sec. (fig. 5), non-ferrous metals, such as copper, brass, bronze, aluminum during 30-40 sec.

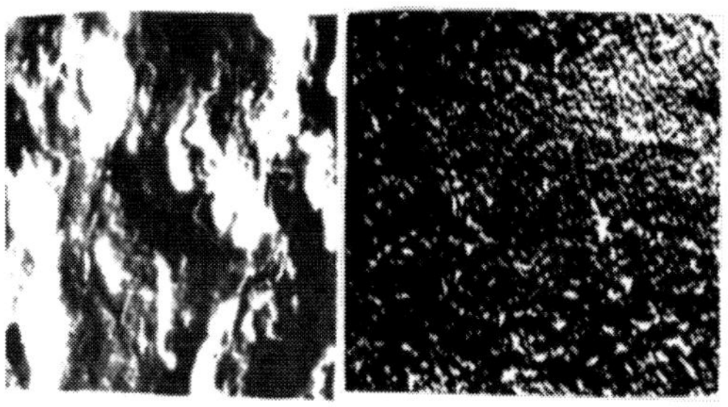

a) b)

Fig. 5. Cleaning of the surface of the semiconductor (germaniye, increase - x 180), time of processing T = 25 sec.

a - the raw surface;

b- the processed surface.

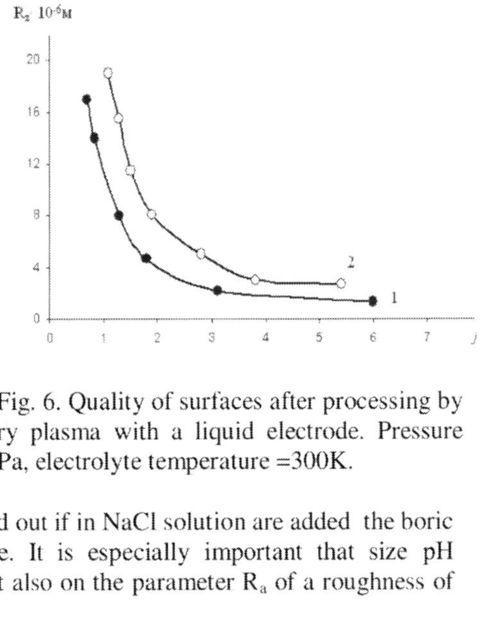

Fig. 6. Quality of surfaces after processing by category plasma with a liquid electrode. Pressure $P=10^5$ Pa, electrolyte temperature =300K.

The most high-quality cleaning of surfaces is carried out if in NaCl solution are added the boric or lemon acids reducing a overalcalinity of electrolyte. It is especially important that size pH influences not only on the anode density of current j_a, but also on the parameter R_a of a roughness of the surface.

Addition of sodium nitrate serve as inhibitor of corrosion and after processing the cleared surface becomes resistant. Category tension (≈ 120 V) practically does not influence on parameters of quality of the surface and the speed of cleaning of a detail. Density of anode current influences both on quality of processing and on processing duration. For example, for 20 sec. of processing of a detail current with a density of 1 A/cm^2 with a depth of immersion of h=0,5 ·10^{-3}, we received a roughness of the surface of the 8th class. The increase of current density to 3 A/cm^2 (at duration of time of processing 15sec) allowed to improve quality of cleaning of the surface still. The surface roughness in accordance with GOST 2789 - 73 corresponded to a class 9-10 [10]. The further increase in density of current or processing time practically does not influence on the class of a roughness of the surface. Temperature of electrolyte influences on quality of the processed surface. Our researches showed that with increase in temperature of electrolyte process productivity decreases. Therefore we cooled electrolyte by means of flowing tap water. In (fig. 6) there are results of researches of dependence of the R_z parameter of quality of the processed surfaces of details from copper M1 (a curve 1 are given at U = 120V and time of processing t = 40sec) and steel chrome moly (25XHBA) (a curve 2, magnetowire steel, U = 90V, t = 60 sec.) at processing by the category with a liquid electrode

(electrolyte 300K temperature) depending on current density in a solid electrode. It is visible that at increase in density of current more than 3÷4 A/cm^2 noticeable improvement of quality of surfaces it is not observed.
Methods of cleaning and fair processing raise a roughness class to 8÷10 [11], and diamond processing [12], polishing by paste, electrochemical processing [13] allow to reach the class of a roughness 13÷14. These methods not always are reliable, they are characterized by the low productivity, labor input and unevenness dross on edges, i.e. process of formation of geometrical characteristics of the surface is unhandy process.

3. Conclusion

The issue mechanism of conductivity of a steam-gas attached anode cover allows to explain the following experimental data and regularities:

- cleaning is reached only in solutions with rather high specific conductivity (over 0,05 Ohm^{-1} · cm^{-1}) as the blanket of electrolyte has to contain necessary number of the ions providing an excess charge.
- limit temperature of cleaning (~ 800°C), excluding thermal reflex of steel products is connected with limited issue ability of boiling solution.
- experimentally observed sedimentation of components of solution on the surface of the anode is connected with issue in a steam-gas cover of ions of both signs with prevalence of the negative. Out of solution the interaction of ions of opposite signs will not be weakened any more by hydration that will lead to crystallization of nonvolatile components of solution. The crystallized connections will go to the atmosphere together with couples, will collect on the walls of a working chamber or will sink on the anode by a continuous layer.

4.References

[1] Smirnov N S, M E Gawks 1978 *cleaning of a surface became* M 230 p
[2] Morozov A P etc. 1985 *Electrodigit cleaning of a katanka.* // Heattechnical questions of application questions NTP in metallurgy//Sverdlovsk,.p 102-104
[3] Abdullin I Sh, Kashapov N F and Kudinov V V 2000 *Fizika i Khimiya Obrabotki Materialov* **3** 35-38
[4] Abdullin I Sh, Kashapov N F and Kudinov V V 2000 *Fizika i Khimiya Obrabotki Materialov* **2** 40-44
[5] Gaysin F M, Dream E E, Shakirov Yu I 1990 *The volume category in the steam-gas environment between firm and liquid electrodes*, Moscow
[6] Shakirov Yu I 1990 Diss. of Cand.Tech.Sci. *Characteristics of plasma electrothermal installation with the liquid cathode* - Leningrad 132 p
[7] Gaysin F M, Shakirov Yu I, Hakimov R G 1990 *Category research between firm and liquid electrodes//*Theses of reports of the Republican scientific and technical KAMAZ-KAMPI conference, Chelny Emb. P 161
[8] Shakirov Yu I 2012 *Multichannel plasma installation with the electrolytic cathode.* Tekhnich magazine "Automotive industry". Moscow No. 2 of p 36-38
[9] Garbarz-Olivier J 1972 *Interpretation de l, effect de cathode dans les solutions aqueouses de quelques electrolytes* Acad. Sc. Paris P 275 P 1359-1362
[10] Lyubomudrov S A 2012 *Metrologiya, standardization and certification: accuracy rationing*: Textbook - M.: Research Center Infra-M 206 p
[11] Long A 2010 *I metalwork works: Education guidanc* - M.: Alpha M: INFRA-M 528 p
[12] Ivanov I S 2009 *Tekhnologiya of mechanical engineering*: Studies. grant - M.: INFRA-M,. - 192 p
[13] Adaskin A M, Zuyev V M 2010 *Materialovedeniye and technology of materials* - M: Forum,.-336 p.

IOP Conf. Series: Materials Science and Engineering **69** (2014) 012044 doi:10.1088/1757-899X/69/1/012044

Method of calculating of optimal grading modes with regard to forming of grinding surface roughness

R R Ziyatdinov
Kazan Federal University,
423800, 68/19 Mira avenue, Naberezhnye Chelny, Russia

E-mail: chelny@kpfu.ru

Abstract. The factor affecting the grinding time and surface quality are considered. The methods of calculating the optimum cutting conditions for the grinding operations are suggested.

1. Introduction

One of the promising directions of development of active control systems is their use in automated systems for surface quality control.

In the grinding process there is a gradual decrease in the size of the grinding wheel and the change in cutting ability. On linear dimensions processing error using active control devices sized circle wear does not affect in many cases. However, the change in cutting ability of the wheel during its resistance causes a change of the cutting forces and hence to the appearance of various power and thermal deformations of the system, thus increasing the roughness of the processed surface [1; 2; 3, 4].

Thus, the calculation of the optimal grinding modes should be based on models that include roughness parameters $R_A = f(X)$ (where X – cutting modes).

2.Restrictions selection

Ground surface roughness height is largely determined by the grinding modes. To a decrease the roughness is needed to increase grinding speed, radial and tangential innings decrease [5; 6; 7; 8; 9]. Dependence of the parameter R_A of the ground surface on the wheel speed v_S, radial feed S_R and tangential feed velocity v_K are shown on the fig.1 [10].

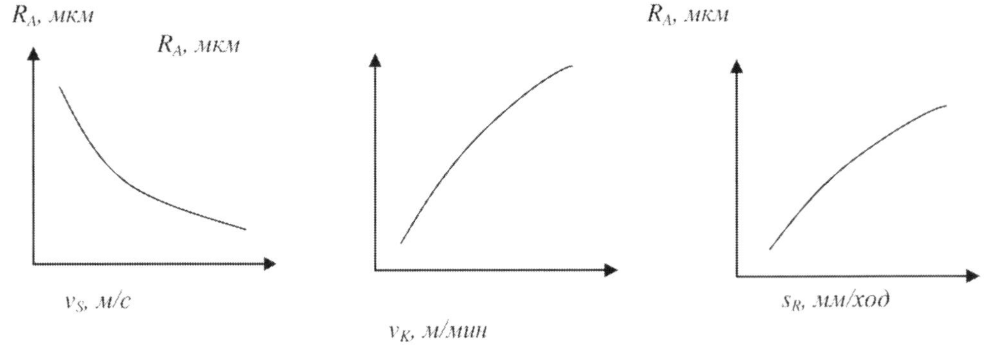

Fig.1. Dependence of the roughness parameter R_A on grinding modes

Ideally, the high processing speed reduces the manufacturing parts time. However, due to the heating of the surface layer may occur the microhardness structure decrease due to the changes (release) of the metal in the surface layer. Such a reduction of microhardness affects the performance properties of the parts adversely.

Grinding burn marks are also rude defects in the surface layer of the metal. These defects are formed when grinding is performed at unacceptably high cutting modes.

The determination of optimal grinding mode is necessary to ensure high performance, the accuracy and quality of the ground surface at the lowest cost of production.

Since cutting modes is closely correlated with the majority of quality processing indicators, the calculation of grinding modes should be carried out taking into account the conditions for obtaining roughness surface no more than a given.

Formulation of any optimization problem begins by defining a set of independent variables, and typically includes conditions that characterize their acceptable values. Thus, in solving the problem of minimizing processing time T_{PR}, is necessary to solve:

$$T_{ОБР} \to \min ,$$

under the constraints:

$$R_A \le R_{AMAX} \tag{1}$$

$$\begin{cases} v_{SMAX} \le v_S \le v_{SMAX} \\ S_{RMIN} \le S_R \le S_{RMAX} \\ v_{KMIN} \le v_K \le v_{KMAX} \end{cases} \tag{2}$$

Criterion restrictions are selected from the following considerations :

- Surface roughness parameters are selected according to the requirements of the design documentation (1);

- Rotation speed v_S circle should be the maximum considering the drive possibilities of the machine , but no more than v_{SMAX}, in which the heating surface can lead to a reduction of parts surface microhardness (dispensing) , as well as the strength of the abrasive tool ;

- Radial feed (allowance) S_R should be greater than the depth of the defective metal layer formed on the previous operations and sufficient to correct the residual error of the part shape ;

- Tangential flow v_K must be maximum to reduce the processing time , but not more than the value at which there are power and thermal deformation of the items, and as a consequence an roughness increase in parts.

In constructing a mathematical model of a particular grinding process is necessary to know the numerical values of the parameters and the patterns of their changes under varying process conditions. Since the grinding process is characterized by great variability of parameters and processing conditions, the reliable data can only be obtained by experimental study of a particular process.

To improve the quality in the existing technology is necessary to carry out experiments to assess the effect of the various factors on surface roughness and conditions using the experimental design method. As the variable factors are taken tangential velocity value v_k, radial feed S_R, tangential flow v_S. For a given process it is possible to take the additional parameters Xi, such as the properties of the abrasive tool , coolants etc. Output parameters are processing time T_{PR} and R_A processed surface roughness .

As an entry-level are selected the technological factors values corresponding to the normal process modes. Variation interval is chosen, on the one hand, several times to be fitted into the range of variation of the variable factor, and on the other hand would allow to ignore the errors of the executive and measurement instruments and mechanisms [11].

On the basis of the experiment a model of surface roughness depending on factors is constructed:

$$R_A = f(v_K, S_R, v_S, X_1, ..., X_m).$$

Substituting into the formula instead of the R_A the maximum allowable roughness inherent in the design documentation and taking into consideration the restrictions (2), we obtain the range of the cutting parameters on the basis of the desired quality parameter.

Taking into consideration the limitations imposed by the machine characteristics (maximum and minimum speed drives, etc.), we obtain the range of the cutting modes parameters change:

$$\begin{cases} v_{SMIND} \le v_S \le v_{SMAXD} \\ S_{RMIND} \le S_R \le S_{RMAXD} \\ v_{KMIND} \le v_K \le v_{KMAXD} \end{cases} \quad (3)$$

Here, the index «D» means allowable cutting conditions based on the required surface roughness R_A and based to the restrictions (2).

Minimization problem. Processing time is determined by three factors: the tangential velocity v_k, radial feed S_R, tangential flow v_S.

Using method of mathematical planning of the experiment we will get a model to calculate the processing time depending on the cutting conditions:

$$T_{OBP} = f(v_K, S_R, v_S)$$

Thus, the problem of minimizing the processing time is reduced to solution:

$$T_{ОБР} = f(v_K, S_R, v_S) \rightarrow \min$$

under the constraints (3).

Most often, the model is described by continuous functions of second or third order [12], and therefore can be applied to optimize the gradient method.

$$gradf(v_K, S_R, v_S) = \left[\frac{\partial f(v_K, S_R, v_S)}{\partial v_K}, \frac{\partial f(v_K, S_R, v_S)}{\partial S_R}, \frac{\partial f(v_K, S_R, v_S)}{\partial v_S} \right]$$

To find the minimum of the function T_{PR} vector g of unit length of search direction is determined:

$$g = -\frac{\overline{gradf}(v_K, S_R, v_S)}{|gradf(v_K, S_R, v_S)|}$$

3.Conclusion. Solution of the problem of minimizing the processing time with the constraints of the desired surface roughness allows you to get the optimum grinding modes, which increases tool life, reduces the processing time and reduces defects output.

4.References

[1] Bezyazichny V F 2013 *Principles of engineering technology: a textbook for high schools.* - M.: Mechanical Engineering

[2] *Active control in mechanical engineering*: / Ed. E I Pedya. 1978 - 2nd ed., Rev. and add. - M.: Mechanical Engineering

[3] Poletaev V A Volkov D I 2009 *Deep grinding of turbine blades - technologist library.* - M.: Mechanical Engineering

[4] Abdullin I Sh, Galyautdinov R T and Kashapov N F 2001 *Engeneers Journal* **74 (5)** 104-107.

[5] *Processing technology by abrasive and diamond tool: Tutorial for engineering colleges* 1989 Z I Flint, G I Butroin, V M Kolomazin etc.: Under ed. Z I Flint. - L.: Mechanical Engineering. Leningrad department

[6]*Influence of cutting conditions in the deployment of holes on the surface roughness of viscous steels* 2003 / Y P Holmogortsev - Journal of Mechanical Engineering № 3

[7] *Influence of parameters on the optimal concentration of diamond in the wheels on organic knits* 1999 / A V Popov - Journal of Mechanical Engineering № 12

[8] Smirnov G A, Bahar V P, Kulikov M Yu, Antyufeeva T P *Investigation of the finish diamond processing processes of corundum ceramics* 2001/ - Bulletin of mechanical engineering, № 9

[9] Abasov V A, Babaev S G, Onalla A I *Surface roughness of the parts from powder new materials at flat grinding* 1999 / - Journal of Mechanical Engineering № 10

IOP Conf. Series: Materials Science and Engineering **69** (2014) 012044 doi:10.1088/1757-899X/69/1/012044

[10] Z I Flint Z I, Butroin G I, V M Kolomazin V M etc *Processing technology by abrasive and diamond tools : Tutorial for engineering college* 1989.; Under ed. Z I Flint. - L.: Mechanical Engineering. Leningrad department
[11] Valkov V M, V E 1973 *Summits Automated process control systems*. L., Mechanical Engineering
[12] Khudobin L V, Guryanikhin V F, Yuganov V S *Diagnosis and management of the grinding at the sound pressure amplitude* 2000/ - Bulletin of mechanical engineering № 1

AUTHOR INDEX

Institute of Physics
Dirac House, Temple Back
Bristol BS1 6BE UK

proceedings.com